TJ
246
.C93
1996

Gaskets

Gaskets

Design, Selection, and Testing

Daniel E. Czernik
V.P. Advanced Technology, FEL-PRO Inc.
Skokie, Illinois

McGraw-Hill

New York San Francisco Washington, D.C. Auckland Bogotá
Caracas Lisbon London Madrid Mexico City Milan
Montreal New Delhi San Juan Singapore
Sydney Tokyo Toronto

Library of Congress Cataloging-in-Publication Data

Czernik, Daniel E.
    Gaskets / Daniel E. Czernik.
       p.   cm.
    Includes index.
    ISBN 0-07-015113-X
    1. Gaskets—Congresses—Handbooks, manuals, etc. I. Title.
TJ246.C93   1996
621.8'85—dc20   95-36623
                                                         CIP

McGraw-Hill
*A Division of The **McGraw·Hill** Companies*

Copyright © 1996 by The McGraw-Hill Companies, Inc. All rights reserved. Printed in the United States of America. Except as permitted under the United States Copyright Act of 1976, no part of this publication may be reproduced or distributed in any form or by any means, or stored in a data base or retrieval system, without the prior written permission of the publisher.

1 2 3 4 5 6 7 8 9 0   QBP/QBP   9 0 1 0 9 8 7 6

ISBN 0-07-015113-X

The sponsoring editor for this book was Robert Hauserman, the editing supervisor was Bernard Onken, and the production supervisor was Donald Schmidt. It was set in Century Schoolbook by Donald A. Feldman of McGraw-Hill's Professional Book Group composition unit.

Printed and bound by Quebecor/Book Press.

McGraw-Hill books are available at special quantity discounts to use as premiums and sales promotions, or for use in corporate training programs. For more information, please write to the Director of Special Sales, McGraw-Hill, 11 West 19th Street, New York, NY 10011. Or contact your local bookstore.

Information contained in this work has been obtained by The McGraw-Hill Companies, Inc. ("McGraw-Hill") from sources believed to be reliable. However, neither McGraw-Hill nor its authors guarantees the accuracy or completeness of any information published herein and neither McGraw-Hill nor its authors shall be responsible for any errors, omissions, or damages arising out of use of this information. This work is published with the understanding that McGraw-Hill and its authors are supplying information, but are not attempting to render engineering or other professional services. If such services are required, the assistance of an appropriate professional should be sought.

    This book is printed on recycled, acid-free paper containing 10% postconsumer waste.

Contents

Preface   ix
Acknowledgments   xi
Introduction   xiii

Chapter 1. Gasket Types   1

Definition   1
Nonmetallic Gaskets   2
Test Procedures for Gasket Materials   7
Gasket Material Manufacturing   25
Nonasbestos Gasketing Materials   30
Important Material Characteristics for Processing and/or Assembly   32
Material Properties That Are Important in Gasketing   34
Fiber Materials   37
Binders   39
Filler Materials   41
Nonfiber Gasket Materials   41
Metallic Gaskets   42
Gasket Types   45

Chapter 2. The Gasket and Its Environment   51

Gasket Design and Environmental Conditions   51
Initial Seal Creation   66
Stress-Distribution Testing   69

Chapter 3. The Gasket and the Joint   89

Maintenance of the Seal   89
Relaxation   90
Effects of Unit Operation   93
Effects of Temperature   95
Gasket Shape Factor   96
Relaxation and Torque Loss   98

Joint and Gasket Design Selection ... 104
Potential New Gasket Code ... 117

Chapter 4. Pressure Vessel Research Council ... 127

PVRC Gasket Testing and Analysis ... 127
PVRC Gasket and Bolted Joint Research Programs ... 128
Mechanical Behavior of Gaskets ... 132
Elevated Temperature Research Program ... 133
Quantification of Aging Effects ... 135
New Quantification Tools ... 137
Qualification Guides or Protocols ... 147
Continuing Research Effort ... 148

Chapter 5. The Gasket and the Application ... 153

Gasket and Joint Diagram ... 153
Application Information ... 158
Gasket Installation ... 158
Sealing Enhancements ... 160
Segmented Gaskets ... 186

Chapter 6. Rubber Gaskets ... 189

Rubber ... 189
Manufacture ... 199
Glossary of Terms Relating to Rubber and Rubberlike Materials ... 203
O-Ring Seals ... 210

Chapter 7. Gasket Testing ... 221

Gasket Material Tests and Their Significance ... 221
Gasket Material Analysis Techniques ... 235
Gasket Material and Gasket Testing ... 240
Bench Testing ... 246

Chapter 8. Gasket Analysis ... 253

Failure Mode and Effects Analysis (FMEA) ... 253
Finite Element Analysis (FEA) ... 260

Chapter 9. Gasket Leakage and Chemical Gaskets ... 265

Leakage—Detection and Rating ... 265
Gasket Fabrication ... 271
Chemical Gaskets ... 276

Chapter 10. Engine Gaskets — 287

Internal Combustion Engine Gaskets — 287
Head Gaskets—Combustion Sealing — 297
Head Gaskets—Liquid Sealing — 302
Intake and Exhaust Manifold Gaskets — 308
Other Gaskets — 312
Chemical Gasketing — 313
Testing of Engine Gaskets — 314

Index — 329

Preface

In the field of gaskets and seals, the gaskets are generally associated with sealing mating flanges of a mechanical joint while seals are generally associated with sealing reciprocating or rotating shafts or moving parts.

Some designers refer to gaskets as static seals and consider seals to be dynamic sealing components. This, however, is not really true since the relationship between a gasket and its mating flanges is dynamic in nature. The reasons for this are the deformations associated with internal pressure, vibration, external forces, and thermal expansion or contraction. These deformations are dynamic.

This book covers gaskets. It includes chemical gaskets and O-rings as well as metallic and nonmetallic gaskets. Seals for rotating shafts and/or moving parts are not included.

Daniel E. Czernik

Acknowledgments

Fel-Pro Inc.—for making it possible.
David South—for his suggestion that a handbook be created.
Leslie A. Horve of CR Industries—for his encouragement.
Don J. McDowell of McDowell Technology—for his aid during its preparation and his review of it.
Fel-Pro Inc.—for providing all of the artwork.
The following specific Fel-Pro employees:
Leticia Ramirez—for typing the manuscript.
Darrell Dimmick, Dawn Hentrich, Kathy Pinto, and Char Ksepka for creating the art.
All the rest of the Fel-Pro people who helped me during this endeavor.

Contributions

Chapter 4 was authored by Michel Derenne[*] and J. R. Payne[†] in collaboration with Luc Marchand[*] and Andre Bazergui.[*]

---

[*]Ecole Polytechnique of Montreal.
[†]Process equipment consultant.

Introduction

Performance, efficiency, reliability, and regard for the environment are important concerns associated with gaskets. They must withstand the higher temperatures and pressures of today's applications and they must have long service life. Respect for the environment is also important not only because of severe governmental regulations but also to ensure the quality of life in the future. This book assesses gasket design relative to these concerns. In addition, the effect of the gasketing environment and the roles of the flanges and fasteners complement the chapters on the various gaskets.

Gasket design and selection associated with creating the initial seal and maintaining the seal are covered. Gasket testing techniques and methods are discussed.

Various methods of stress-distribution testing are illustrated and leakage rating and detection techniques are identified. Use of the computer in gasket selection and finite element analysis of gasket designs are introduced.

The current American Society of Mechanical Engineers (ASME) code and the proposed new ASME gasket constants, which will likely be an addendum to the code, bring the reader up to date in this regard.

Chapter

1

Gasket Types

Definition

Webster's New Collegiate Dictionary defines a gasket as plaited hemp or tallowed rope for packing pistons, making pipe joints, etc., hence, packing or any other suitable material.

The *American Heritage Dictionary* defines a gasket as any of a wide variety of seals or packings used between matched machine parts or around pipe joints to prevent the escape of a gas or fluid.

The American Society for Testing and Materials (ASTM) defines a gasket as a material which may be clamped between faces and acts as a static seal. Gaskets may be cut, formed, or molded to the desired configuration.

Another definition and one this handbook prefers is: A gasket is a material or combination of materials clamped between two separable members of a mechanical joint. Its function is to effect a seal between the members (flanges) and maintain the seal for a prolonged period of time. The gasket must be capable of sealing the mating surfaces, impervious and resistant to the medium being sealed, and able to withstand the application temperature and pressure. Figure 1.1 depicts the nomenclature associated with a gasketed joint.

The gasket and the joint must be considered together. The gasket may work or fail according to whether the joint is designed in accordance with the properties of the gasket itself.

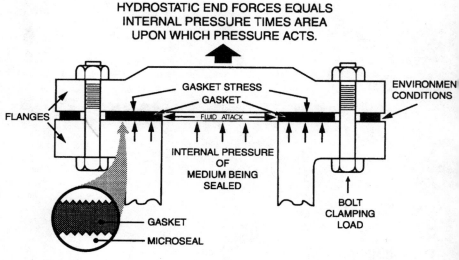

Figure 1.1 Nomenclature of a gasketed joint.

Therefore, the system must be considered as a whole to determine sealing performance. The effect on sealing performance by the joint design including the fasteners is discussed later.

The range of environmental conditions for gaskets is extremely large. Some applications with low clamp load, along with thin flanges such as those associated with dishwasher water pumps, are on the one extreme. On the other extreme are the rigid joints with high clamp load associated with American Society of Mechanical Engineers (ASME) code flanges. The code specification, while being applicable for the latter, is not useful for the wide variety of industrial gasketing, which do not have uniform flange thicknesses, bolt sizes and spacings, etc. This handbook contains information for the gasket engineer to consider when designing gaskets for joints where the ASME code is usable and also for joints when the code is not applicable.

Nonmetallic Gaskets

Most nonmetallic materials consist of a fibrous base with various fillers and miscellaneous chemicals held together or strengthened with an elastomeric binder. The choice of combination of binder, fillers, and base depends upon the media being sealed and the conditions of the sealing environment as well as the load-bearing requirements of the application. Natural or synthetic rubbers, graphite, and plastics are also used for gasketing.

Another category of gaskets is sealants. These are generally high-viscosity liquids that cure in place after the joints have been assembled. They generally are called formed-in-place gaskets. These types of gaskets are discussed later.

As mentioned, gaskets are used to seal joints. There are, however, a number of other requirements of gaskets depending upon the applications.

Requirements of gaskets

- Heat and media resistance
- Zero leakage through the gasket
- Zero leakage over the gasket
- Be affordable
- Be environmentally safe
- Accommodate surface finish conditions of flanges
- Reduce and/or control port distortion
- Reduce and/or control flange distortion
- Accommodate thermal expansion and contraction
- Possess adequate recovery
- Minimize torque loss
- No retorque
- Transfer heat as desired
- Close tolerance on compressed thickness—maintain shim thickness
- Meter fluid
- Provide acoustic or thermal isolation
- Have antistick properties

- Pass customer verification testing
- Manufacturability and handling—robotics, cleanliness, block, visibility for sensors, fit
- Accommodate service assembly requirements
- Be recyclable

Standard classification system for nonmetallic gasket materials

The American Society for Testing and Materials (ASTM) F104 classification system provides a means for specifying or describing pertinent properties of commercial nonmetallic gasket materials. Materials composed of asbestos, cork, cellulose, and other organic or inorganic materials in combination with various binders or impregnants are included. It should be noted that although asbestos is listed in the ASTM classification system, it has been essentially eliminated from the vast majority of gasketing materials. Rubber compounds are not included, since they are covered in ASTM Method D2000 (SAE J200). Also, gasket coatings are not covered since coating details and specifications are intended to be given on engineering drawings or in separate documents.

The ASTM classification is based on the principle that nonmetallic gasket materials can be described in terms of specific physical and mechanical characteristics. Thus users of gasket materials can, by selecting different combinations of statements, specify different combinations of properties desired in their gaskets. Suppliers, likewise, can report properties available in their products in accordance with this classification system.

In specifying or describing gasket materials, each line callout includes the number of the system (minus date symbol) followed by the letter F and six numerals, for example, ASTM F 104 (F125400). Each numeral of the call-out represents a characteristic, and six numerals are always required. The numeral 0 is used when the description of any characteristic is not desired. The numeral 9 is used when the description of any characteristic (or related test) is specified by some supplement to this classification system, such as notes on engineering drawings.

Gasket Types

ASTM F104 Nonmetallic Gasket Classification

Basic six-digit number	Basic characteristic
First number	Type of material (the principal fibrous or particulate reinforcement material from which the gasket is made) shall conform to the first number of the basic six-digit number as follows:

                            0 = not specified
                            1 = asbestos
                            2 = cork
                            3 = cellulose
                            4 = fluorocarbon polymer
                            5 = flexible graphite
                            7 = nonasbestos, tested as type 1
                            9 = as specified*

Second number   Class of material (method of manufacture or common trade designation) shall conform to the second number of the basic six-digit number as follows:

When first number is 1 or 7, for second number
0 = not specified
1 = compressed sheeter process
2 = beater addition process
3 = paper and millboard
9 = as specified*

When first number is 2, for second number
0 = not specified
1 = cork composition
2 = cork and elastomeric
3 = cork and cellular rubber
9 = as specified*

When first number is 3, for second number
0 = not specified
1 = untreated fiber—tag, chipboard, vulcanized fiber, etc.
2 = protein treated
3 = elastomeric treated
4 = thermosetting resin treated
9 = as specified*

When first number is 4, for second number
0 = not specified
1 = sheet PTFE
2 = PTFE of expanded structure
3 = PTFE filaments, braided or woven
4 = PTFE felts
5 = filled PTFE
9 = as specified*

ASTM F104 Nonmetallic Gasket Classification (*Cont.*)

Basic six-digit number	Basic characteristic
When first number is 5, for second number	0 = not specified 1 = homogeneous sheet 2 = laminated sheet 9 = as specified*
Third number	Compressibility characteristics, determined in accordance with Test Method F36, shall conform to the percentage indicated by the third numeral of the basic six-digit number (example: 4 = 15 to 25% maximum): 0 = not specified 5 = 20 to 30% 1 = 0 to 10% 6 = 25 to 40% 2 = 5 to 15%† 7 = 30 to 50% 3 = 10 to 20% 8 = 40 to 60% 4 = 15 to 25% 9 = as specified*
Fourth number	Thickness increase when immersed in ASTM no. 3 oil, determined in accordance with Test Method F146, shall conform to the percentage indicated by the fourth numeral of the basic six-digit number (example: 4 = 15 to 30%): 0 = not specified 5 = 20 to 40% 1 = 0 to 15% 6 = 30 to 50% 2 = 5 to 20% 7 = 40 to 60% 3 = 10 to 25% 8 = 50 to 70% 4 = 15 to 30% 9 = as specified*
Fifth number	Weight increase when immersed in ASTM no. 3 oil, determined in accordance with Test Method F146, shall conform to the percentage indicated by the fifth numeral of the basic six-digit number (example: 4 = 30% maximum): 0 = not specified 5 = 40% max. 1 = 10% max. 6 = 60% max. 2 = 15% max. 7 = 80% max. 3 = 20% max. 8 = 100% max. 4 = 30% max. 9 = as specified*
Sixth number	Weight increase when immersed in water, determined in accordance with Test Method F146, shall conform to the percentage indicated by the sixth numeral of the basic six-digit number (example: 4 = 30% maximum): 0 = not specified 5 = 40% max. 1 = 10% max. 6 = 60% max. 2 = 15% max. 7 = 80% max. 3 = 20% max. 8 = 100% max. 4 = 30% max. 9 = as specified*

ASTM F104 Nonmetallic Gasket Classification (*Cont.*)

Basic six-digit number	Basic characteristic

*On engineering drawings or other supplement to this classification system. Suppliers of gasket materials should be contacted to find out what line call-out materials are available. Refer to ANSI/ASTM F 104 for further details.

†From 7 to 17% for compressed sheeter process.

A compilation of this classification system in one table is depicted in Table 1.1.

In addition to the classification system, the American Society for Testing and Materials (ASTM) and the Material Test Institute/Pressure Vessel Research Institute (MTI and PVRI) associated with the American Society of Mechanical Engineers (ASME) have identified additional significant tests that can be used to predict a gasket material's performance. These test specifications follow.

Test Procedures for Gasket Materials

American Society for Testing and Materials (ASTM) standard test methods

The following listing identifies the ASTM Standard Test Methods for gasket materials. Following the listing are the scopes for each of the procedures. The ASTM is located at 1916 Race Street in Philadelphia, PA 19103.

F36	Test Method for Compressibility and Recovery of Gasket Materials
F37	Test Method for Sealability of Gasket Materials
F38	Test Method for Creep Relaxation of a Gasket Material
F112	Test Method for Sealability of Enveloped Gaskets
F145	Recommended Practice for Evaluating Flat-Face Gasketed Joint Assemblies
F146	Test Method for Fluid Resistance of Gasket Materials
F147	Test Method for Flexibility of Nonmetallic Gasket Materials Containing Asbestos or Cork
F148	Test Method for Binder Durability of Cork Composition Gasket Materials

TABLE 1.1 Basic Physical and Mechanical Characteristics for Nonmetallic Gasket Materials

Basic six-digit number	Basic characteristics								
	0	1	2	3	4	5	6	7	8	9
First number, type of material	NS	Asbestos	Cork	Cellulose	Fluoro-carbon polymer	Flexible graphite		Non-asbestos		AS
Second number, class of material when first number is 1 or 7	NS	Compressed sheeter process	Beater process	Paper and millboard						AS
Second number when first number is 2	NS	Cork composition	Cork and elastomer	Cork and cellular rubber						AS
Second number when first number is 3	NS	Untreated fiber	Protein treated fiber	Elastomeric treated fiber	Thermoset resin treated					AS
Second number when first number is 4	NS	Sheet PTFE	PTFE with expanded structure	PTFE filaments braided or woven	PTFE felts	Filled PTFE				AS
Second number when first number is 5	NS	Homogeneous sheet	Laminated sheet							AS
Third number, compressibility, ASTM F36, % compression loss	NS	0–10	5–15	10–20	15–25	20–30	25–40	30–50	40–60	AS

Fourth number, % thickness increase when immersed in ASTM no. 3 oil, ASTM F146	NS	0–15	5–20	10–25	15–30	20–40	30–50	40–60	50–70	AS
Fifth number, maximum % weight increase when immersed in ASTM no. 3 oil, ASTM F146	NS	10	15	20	30	40	60	80	100	AS
Sixth number, maximum % weight increase when immersed in water, ASTM F146	NS	10	15	20	30	40	60	80	100	AS

NS = not specified

AS = as specified

F152 Method for Tension Testing of Nonmetallic Gasket Materials
F363 Method for Corrosion Testing of Enveloped Gaskets
F433 Recommended Practice for Evaluating Thermal Conductivity of Gasket Materials
F434 Method for Blowout Testing of Preformed Gaskets
F495 Test Method for Ignition Loss of Gasket Materials Containing Inorganic Substances
F586 Test Method for Leak Rates versus y Stresses and m Factors for Gaskets
F607 Test Method for Adhesion of Gasket Materials to Metal Surfaces
F806 Test Method for Compressibility of Laminated Composite Gasket Materials
F1087 Linear Dimensional Stability of a Gasket Material to Moisture

ASTM test methods for vulcanized elastomers

D395 Compression Set, Method B
D412 Tension Properties
D429 Adhesion Bond Strength, Method A
D430 Flex Resistance
D471 Fluid Resistance—Aqueous, Fuels, Oils, and Lubricants
D573 Heat Age or Heat Resistance
D575 Compression Deflection, Method A
D624 Tear Strength, Die B or C
D813 Crack Growth
D865 Deterioration by Heating
D945 Resilience
D2632 Resilience
D1053 Low-Temperature Torsional
D1171 Ozone or Weather Resistance
D1329 Low-Temperature Retraction
D1418 Elastomer Classification
D2137 Low-Temperature Resistance A
D2240 Hardness Durometer A

Standard classifications

ASTM F104

Classification System for Nonmetallic Gasket Materials. Scope: This classification system provides a means for specifying or describing pertinent properties of commercial nonmetal gasket materials. Materials composed of asbestos, cork, cellulose, and other organic or inorganic materials in combination with various binders or impregnants are included. Materials normally classified as rubber compounds are not included, since they are covered in Method D2000. Gasket coatings are not covered, since details thereof are intended to be given on engineering drawings or in separate specifications.

Since all the properties that contribute to gasket performance are not included, use of the classification system as a basis for selecting materials is limited.

ASTM F868

Classification System for Laminated Composite Gasket Materials. Scope: This classification system provides a means for specifying or describing pertinent properties of commercial laminate composite gasket materials (LCGM). These structures are composed of two or more chemically different layers of material. These materials may be organic or inorganic, or combinations with various binders or impregnants. Gasket coatings are not covered since details thereof are intended to be given on engineering drawings or as separate specifications. Commercial materials designated as envelope gaskets are excluded from this standard.

Since all properties that contribute to gasket performance are not included, use of this classification system as a basis for selecting LCGM is limited.

Standard test methods

ASTM F36

Test Method for Compressibility and Recovery of Gasket Materials. Scope: This test method covers determination of the short-time compressibility and recovery at room temperature of sheet-gasket material and, in certain cases, gaskets cut from sheets. It is not intended as a test for compressibility under prolonged stress application, generally referred to as "creep," or for recovery following such prolonged stress application, the inverse of which is generally referred to as "compression sets." Also, it is not intended for tests at other than room temperature.

ASTM F37

Test Method for Sealability of Gasket Materials. Scope: These test methods provide a means of evaluating fluid sealing properties of gasket materials at room temperature. Method A is restricted to liquid measurements and Method B may be used for both gas and liquid leakage measurements.

These methods are suitable for evaluating the seal characteristics of a gasket material under differing compressive flange load. When desired, the method may be used as an acceptance test when test conditions are agreed upon between a supplier and purchaser as follows: fluid, internal pressure on fluid, and flange load on gasket specimen.

ASTM F38

Test Method for Creep Relaxation of a Gasket Material. Scope: These test methods provide a means of measuring the amount of creep relaxation of a gasket material at a stated time after a compressive stress has been applied.

Method A: Creep relaxation measured by means of a calibrated strain gauge on a bolt.

Method B: Creep relaxation measured by means of a calibrated bolt with dial indicator.

ASTM F112

Test Method for Sealability of Enveloped Gaskets. Scope: This test method covers the evaluation of the sealing properties of enveloped gaskets for use with corrosion-resistant process equipment. Enveloped gaskets are described as gaskets having some corrosion-resistant covering over the internal area normally exposed to the corrosive environment. The shield material may be plastic (such as polytetrafluoroethylene) or metal (such as tantalum). A resilient conformable filler is usually used inside the envelope.

ASTM F145

Recommended Practice for Evaluating Flat-Face Gasketed Joint Assemblies. Scope: This practice permits measurement of gasket compression resulting from bolt loading on a flat-face joint assembly at ambient conditions.

ASTM F146

Test Method for Fluid Resistance of Gasket Materials. Scope: These test methods cover the determination of the effect on physical properties of nonmetallic gasketing materials after immersion in test fluids. The types of materials covered are those containing asbestos and other inorganic fibers (type 1), cork (type 2), and cellulose or other organic fiber (type 3) as described in Classification F104. These methods are not applicable to the testing of vulcanized rubber, a method that is described in Test Method D471. It is designed for testing specimens cut from gasketing materials or from finished articles of commerce.

This standard may involve hazardous materials, operations, and equipment. This standard does not purport to address all the safety problems associated with its use. It is the responsibility of whoever uses this standard to consult and establish appropriate safety and health practices and determine the applicability of regulatory limitations prior to use.

ASTM F147

Test Method for Flexibility of Nonmetallic Gasket Materials Containing Asbestos or Cork. Scope: This test method covers the determination of the flexibility of nonmetallic gasket materials. It is designed for testing specimens cut from sheet goods or from the gasket in the finished form, as supplied for commercial use. Materials normally classified as elastomeric compounds are excluded since they are covered in Classification D2000.

ASTM F148

Test Method for Binder Durability of Cork Composition Gasket Materials. Scope: This test method covers three procedures for determination of the binder durability of cork-containing materials.

This standard may involve hazardous materials, operations, and equipment. This standard does not purport to address all the safety problems associated with its use. It is the responsibility of whoever uses this standard to consult and establish appropriate safety and health practices and determine the applicability of regulatory limitations prior to use.

ASTM F152

Method for Tension Testing of Nonmetallic Gasket Materials. Scope: These test methods cover the determination of tensile strength of certain nonmetallic gasketing materials at room temperature. The types of materials covered are those pertaining to asbestos and other inorganic fibers (type 1), cork (type 2), and cellulose or other organic fiber (type 3) as described in Method F104. These methods are not applicable to the testing of vulcanized rubber (see Method D412) or rubber O-rings (see Method D1414).

ASTM F363

Method for Corrosion Testing of Enveloped Gaskets. Scope: This test method covers the evaluation of gaskets under corrosive conditions at varying temperature and pressure levels. The test unit may be glass-lined if the flanges are sufficiently plane (industry accepted), thus providing resistance to all chemicals, except hydrofluoric acid, from cryogenic temperatures to 260°C (500°F) at pressures from full vacuum to the allowable pressure rating of the unit, or made of other suitable material. The test unit described has an internal design pressure rating of 1034 kPa (150 psi) at 260°C.

ASTM F433

Recommended Practice for Evaluating Thermal Conductivity of Gasket Materials. Scope: This practice covers a means of measuring the amount of heat transfer quantitatively through a material or system. This recommended practice is similar to the heat flowmeter system of Method C518 but modified to accommodate small test samples of higher thermal conductance.

ASTM F434

Method for Blowout Testing of Preformed Gaskets. Scope:

This test method covers the determination of the resistance against blowout of preformed gaskets. The test is conducted under ambient conditions and should be used for comparison purposes only to select suitable designs and constructions for specific applications.

This standard may involve hazardous materials, operations, and equipment. This standard does not purport to address all the safety problems associated with its use. It is the responsibility of whoever uses this standard to consult and establish appropriate safety and health practices and determine the applicability of regulatory limitations prior to use.

ASTM F495
Test Method for Ignition Loss of Gasket Materials Containing Inorganic Substances. Scope: This test method covers the determination of gasket material weight loss upon exposure to elevated temperatures.

ASTM F586
Test Method for Leak Rates versus y Stresses and m Factors for Gaskets. Scope: This test method covers the determination of leak rates versus y stresses and m factors for gaskets gripped by pressure-containing flanged connections.

ASTM F607
Test Method for Adhesion of Gasket Materials to Metal Surfaces. Scope: This test method provides a means of determining the degree to which gasket materials under compressive load adhere to metal surfaces.

The test method may be employed for the determination of adhesion. The test conditions described are indicative of those frequently encountered in gasket application. Test conditions may also be modified in accordance with the needs of specific applications as agreed upon between the user and the producer.

This standard may involve hazardous materials, operations, and equipment. This standard does not purport to address all the safety problems associated with its use. It is the responsibility of whoever uses this standard to consult and establish appropriate safety and health practices and determine the applicability of regulatory limitations prior to use.

ASTM F806

Test Method for Compressibility and Recovery of Laminated Composite Gasket Materials. Scope: This test method covers determination of the short-term compressibility and recovery at room temperature of laminated composite gasket materials.

This test method is not intended as a test for compressibility under prolonged stress application, that is, "creep," or for recovery following such prolonged stress application, the inverse of which is generally referred to as "compression set." Also, it is only intended for tests at room temperature.

This standard may involve hazardous materials, operations, and equipment. This standard does not purport to address all the safety problems associated with its use. It is the responsibility of whoever uses this standard to consult and establish appropriate safety and health practices and determine the applicability of regulatory limitations prior to use.

ASTM F1087

Linear Dimensional Stability of a Gasket Material to Moisture. Scope: This method covers a procedure to determine the stability of a gasket material to linear dimensional change due to hydroscopic expansion and contraction. It subjects a sample to extremes, i.e., oven drying and complete immersion in water, which have shown good correlation to low and high relative humidities.

ASTM test methods for vulcanized elastomers

ASTM D395

Compression Set, Method B. Scope: These test methods cover the testing of rubber intended for use in applications in which the rubber will be subjected to compressive stresses in air or liquid media. They are applicable particularly to the rubber used in machinery mountings, vibration dampers, and seals. Two methods are covered as follows:

Method A: Compression Set under Constant Force in Air, sec. 7.10

Method B: Compression Set under Constant Deflection in Air, sec. 11.14

The choice of method is optional, but consideration should be given to the nature of the service for which correlation of test results may be sought. Unless otherwise stated in a detailed specification, Method B should be used.

Method B is not suitable for vulcanizates harder that 90 IRHD.

The values stated in SI units are to be regarded as the standard.

This standard may involve hazardous materials, operations, and equipment. This standard does not purport to address all the safety problems associated with its use. It is the responsibility of whoever uses this standard to consult and establish appropriate safety and health practices and determine the applicability of regulatory limitations prior to use.

ASTM D412

Tension Properties. Scope: These test methods cover tension testing of rubber at various temperatures. These methods are not applicable to the testing of ebonite and similar hard, low-elongation materials. The methods appear as follows:

Test Method A: Dumbbells and Straight Sections, secs. 13 to 16

Test Method B: Cut Ring Specimens, secs. 17 to 20

The standard is used to determine tensile stress, tensile strength, and yield point.

The values stated in either SI or non-SI units should be regarded separately as standard. The values in each system may not be exact equivalents; therefore, each system must be used independently of the other, without combining values in any way.

This standard may involve hazardous materials, operations, and equipment. This standard does not purport to address all the safety problems associated with its use. It is the responsibility of whoever uses this standard to consult and establish appropriate safety and health practices and determine the applicability of regulatory limitations prior to use.

ASTM D429

Adhesion Bond Strength, Method A. Scope: These test methods cover procedures for testing the static adhesional strength of rubber to rigid materials (in most cases metals).

Method A: Rubber Part Assembled between Two Parallel Metal Plates

Method B: 90° Stripping Test—Rubber Part Assembled to One Metal Plate

Method C: Measuring Adhesion of Rubber to Metal with a Conical Specimen

Method D: Adhesion Test—Postvulcanization (PV) Bonding of Rubber to Metal

Method E: 90° Stripping Test—Rubber Tank Lining— Assembled to One Metal Plate

While the method may be used with a wide variety of rigid materials, such materials are the exception rather than the rule. For this reason, we have used the word "metal" in the text rather than "rigid materials."

ASTM D430

Flex Resistance. Scope: These test methods cover testing procedures that estimate the ability of soft rubber compounds to resist dynamic fatigue. No exact correlation between these test results and service is given or implied. This is due to the varied nature of service conditions. These test procedures do yield data that can be used for the comparative evaluation of rubber compounds or composite rubber-fabric materials for their ability to resist dynamic fatigue.

ASTM D471

Fluid Resistance—Aqueous, Fuels, Oils, and Lubricants. Scope: This test method measures the comparative ability of rubber and rubberlike composition to withstand the effect of liquids. It is designed for testing specimens of elastomeric vulcanizates cut from standard sheets (see Recommended Practice D3182), specimens cut from fabric coated with elastomeric vulcanizates (see Method D751), or finished articles of commerce (see Practice D3183). The method is not applicable to the testing of cellular rubbers, porous compositions, and compressed asbestos sheet except as provided in Note 5.

In view of the wide variations often present in service conditions, this accelerated test may not give any direct correlation with service performance. However, the method yields comparative data on which to base judgment as to expected service quality and is especially useful in research and development work.

ASTM D573

Heat Age or Heat Resistance. Scope: This test method describes a procedure to determine the influence of elevated temperature on the physical properties of vulcanized rubber. The results of this test may not give an exact correlation with service performance since performance conditions vary widely. The test may, however, be used to evaluate rubber compounds on a laboratory comparison basis.

ASTM D575

Compression Deflection, Method A. Scope: These test methods describe two test procedures for determining the compression-deflection characteristics of rubber compounds other than those usually classified as hard rubber and sponge rubber.

ASTM D624

Tear Strength, Die B or C. Scope: This test method covers the determination of the tear resistance of vulcanized rubber. It does not apply to testing of hard rubber.

The values stated in SI units are to be regarded as the standard.

This standard may involve hazardous materials, operations and equipment. This standard does not purport to address all the safety problems associated with its use. It is the responsibility of whoever uses this standard to consult and establish appropriate safety and health practices and determine the applicability of regulatory limitations prior to use.

ASTM D813

Crack Growth. Scope: This test method covers the determination of crack growth of vulcanized rubber when subjected to repeated bend flexing. It is particularly applicable to tests of synthetic rubber compounds which resist the initiation of cracking due to flexing when tested by Method B or Method D430. Cracking initiated in these materials by small cuts or tears in service may rapidly increase in size and progress owing to complete failure even though the material is extremely resistant to the original flexing-fatigue cracking. Because of this characteristic of synthetic compounds, particularly those of the SBR type, this method in which the specimens are first artificially punctured in the flex area should be used in evaluating the fatigue-cracking properties of this class of material.

ASTM D865

Deterioration by Heating. Scope: This test method describes a procedure to determine the deterioration induced by heating rubber specimens in individual test tube enclosures with circulating air. This isolation prevents cross contamination of compounds due to loss of volatile materials (for example, antioxidants) and their subsequent migration into other rubber compounds (specimens). The absorption of such volatile materials may influence the degradation rate of rubber compounds.

ASTM D945

Resilience. Scope: These test methods cover the use of the Yerrzley mechanical oscillograph for measuring mechanical properties of elastomeric vulcanizates in the generally small range of deformation that characterizes many technical applications. These properties include resilience, dynamic modulus, static modulus, kinetic energy, creep, and set under a given dead load. Measurements in compression and shear are described.

The test is applicable primarily, but not exclusively, to materials having static moduli at the test temperature such that loads below 2 MPa (280 psi) in compression or 1 MPa (140 psi) in shear will produce 20 percent deformation, and having resilience such that at least three complete cycles are produced when obtaining the damped oscillatory curve. The range may be extended, however, by use of supplementary masses and refined methods of analysis. Materials may be compared under either comparable mean stress or mean strain conditions.

ASTM D2632

Resilience. Scope: This test method covers the determination of impact resilience of solid rubber from measurement of the vertical rebound of a dropped mass.

The method is not applicable to the testing of cellular rubbers or coated fabrics.

NOTE 1-A: Standard method of test for impact resilience and penetration of rubber by a rebound pendulum is described in ASTM Method D1054, Test for Rubber Property Resilience Using a Rebound Pendulum.

ASTM D1053

Low-Temperature Torsional. Scope: These test methods describe the use of a torsional apparatus for measuring the relative low-temperature stiffening of flexible polymeric materials and fabrics coated therewith. A routine inspection and acceptance procedure, to be used as a pass-fail test at a specified temperature, is also described.

These test methods yield comparative data to access the low-temperature performance of flexible polymers and fabrics coated therewith.

The values stated in either SI or non-SI units should be regarded separately as the standard. The values in each system may not be exact equivalents; therefore, each system must be used independently of the other, without combining values in any way.

This standard may involve hazardous materials, operations, and equipment. This standard does not purport to address all the safety problems associated with its use. It is the responsibility of whoever uses this standard to consult and establish appropriate safety and health practices and determine the applicability of regulatory limitations prior to use.

ASTM D1171

Ozone or Weather Resistance. Scope: This test method permits the estimation of the relative ability of rubber compounds used for applications requiring resistance to outdoor weathering or ozone chamber testing.

This test method is not applicable to materials ordinarily classed as hard rubber but is adaptable to molded or extruded soft rubber material and sponge rubber for use in window weather stripping as well as similar automotive applications.

This standard may involve hazardous materials, operations, and equipment. This standard does not purport to address all the safety problems associated with its use. It is the responsibility of whoever uses this standard to consult and establish appropriate safety and health practices and determine the applicability of regulatory limitations prior to use.

ASTM D1329

Low-Temperature Retraction. Scope: This test method covers a temperature-retraction procedure for rapid evaluation of crystallization effects and for comparing viscoelastic properties of rubber and rubberlike materials at low temperatures. This test is useful when employed in conjunction with other low-temperature tests for selection for materials suitable for low-temperature service. It is also of value in connection with research and development but is not yet considered sufficiently well established for use in purchase specification.

ASTM D1418

Elastomer Classification. Scope: This practice establishes a system of general classification for the basic rubbers in both dry and latex forms determined from the chemical composition of the polymer chain.

The purpose of this practice is to provide a standardization of terms for use in industry, commerce, and government; it is not intended to conflict with but rather to act as a supplement to existing trade names and trademarks.

In technical papers or presentations the name of the polymer should be used if possible. The symbols can follow the chemical name for use in later references.

This standard may involve hazardous materials, operations and equipment. This standard does not purport to address all the safety problems associated with its use. It is the responsibility of whoever uses this standard to consult and establish appropriate safety and health practices and determine the applicability of regulatory limitations prior to use.

ASTM D2137

Low-Temperature Resistance A. Scope: These test methods cover the determination of the lowest temperature at which rubber vulcanizates and rubber-coated fabrics will not exhibit fractures or coating cracks when subjected to specified impact conditions.

ASTM D2240

Hardness Durometer A. Scope: This test method covers two types of durometers, A and D, and the procedure for determining the indentation hardness of homogeneous materials ranging from soft vulcanized rubber to some rigid plastics.

This test method is not applicable to the testing of coated fabrics.

The values stated in SI units are to be regarded as the standard.

This standard may involve hazardous materials, operations, and equipment. This standard does not purport to address all the safety problems associated with its use. It is the responsibility of whoever uses this standard to consult and establish appropriate safety and health practices and determine the applicability of regulatory limitations prior to use.

Materials Technology Institute of the Chemical Process Industries (MTI) and the Pressure Vessel Research Council (PVRC)

Listed below are gasket tests that were developed by the MTI and PVRC. Each of these PVRC/MTI gasket tests has an established testing procedure. This guarantees consistency of results for comparison purposes. Several of these procedures are being reviewed by ASTM for establishing consistent, uniform national gasket test procedures. Similar testing is being considered in Europe as a part of ISO standards.

(1) FIRS — Fire Simulation Screen Test: Test procedure utilizing ATRS fixture without springs to obtain relaxation and posttest tensile properties of specimens typically exposed to 1200°F for 20 min

(2) FITT — Fire Simulation Tightness Test: Test procedure utilizing the single gasket 1200°F HOTT fixture to obtain tightness properties of specimens typically exposed to 1200°F for 20 min

(3) ATRS — Aged Tensile/Relaxation Screen Test: Test procedure utilizing spring-loaded fixture to obtain relaxation and posttest tensile properties and weight loss. Temperature to 750°F with aging to 1000 h (or more)

(4) ARLA — Aged Relaxation Leakage Adhesion Test: A screening test procedure utilizing a spring-loaded fixture with ASTM F38 size platens to obtain relaxation and posttest tightness properties, adhesion, and weight loss. Temperature to 750°F with aging to 1000 h (or more)

(5) HATR — High-Temperature Aged Tensile/Relaxation Screen Test: The same procedure as ATRS but fixture extended to 1050°F via improved materials of construction.

(6) ROTT — Room Temperature Tightness Tests

Three pressure levels are used in seating sequences so as to determine the correlation between pressure and leakage. Gasket tightness parameters are obtained for ASME code design use.

(7) HOTT — Hot Operational Tightness Test: Test procedure utilizing a hydraulic fixture to evaluate and confirm the sealing performance of gaskets at elevated temperature. Pressure and temperature are maintained throughout test, which may typically last 5 to 14 days while load, leak rate, and displacement are monitored. Double gasket fixture tests are to 800°F; single gasket fixture tests are to 1200°F. Two test temperatures are required: (a) maximum vendor recommended usage temperature (T_{max}), and (b) ½ the maximum vendor recommended usage temperature ($T_{max}/2$).

(8) EHOTT — This test is a combination of the ROTT and HOTT tests and should be used anytime both of these tests are needed unless you are performing multiple temperature tests, in which case you should use the HOTT tests after running either one EHOTT or one ROTT test.

(9) AHOT — Aged Hot Tightness Test

This leak performance test is conducted at temperatures up to 800°F on pairs of NPS 4-in gaskets after they have been oven aged in special platens for several weeks at temperatures to 1200°F. The AHOT test is used when the exposure times and temperatures are impracticable for the HOTT test.

NPS 4-in gaskets are tested in a three-part sequence consisting of initial compression, aged, and final leak testing. Mounted in special platens, the gaskets are hot compressed at temperatures up to 1200°F in the short-term HOTT fixture and then aged in a still air oven. During aging, the gasket is maintained in an air purge stream or with helium or nitrogen under light pressure.

The platens are transferred to the HOTT apparatus and tightness tested for 2 days at 800°F. This allows a practical evaluation of response to blowout conditions and controlled thermal disturbances. Test duration is 2 to 6 weeks and can be longer if desired.

Facilities that perform all or some of the PVRC/MTI gasket tests are:

Mr. Michel Derenne or
Dr. L. Marchand
Ecole Polytechnic Institute
P.O. Box 6079, Sta. A
Montreal, QE H3C 3A7
  CANADA
Attn.: Fax: 40-37-36-99

Dr. Y. Birembaut
CETIM, Nantes
NANTES CEDEX FRANCE
Phone: 40-74-03-38

Dr. Benard Nau
  BHRG Group Ltd.
Cranfield
Bedford MK43 OAJ England
Phone: 0234-750422
Fax: 9234-750074

Gasket Material Manufacturing

Sheet materials for gaskets are made by four basic processes:

1. Beater addition (fourdrinier and cylinder machines)
2. Compressed (calendering)
3. Graphite (exfoliated and calendering)
4. Reinforced (calendering and combining)

Beater addition. Figure 1.2 depicts a simplified flow diagram of a papermaking machine. This is the basic beater addition process, and the machine is called a fourdrinier. The material ingredients consisting of water, rubber, fillers, fibers, etc., go into the machine, are placed in the machine chest, and are then mixed and put into the head box. A felt belt goes through the

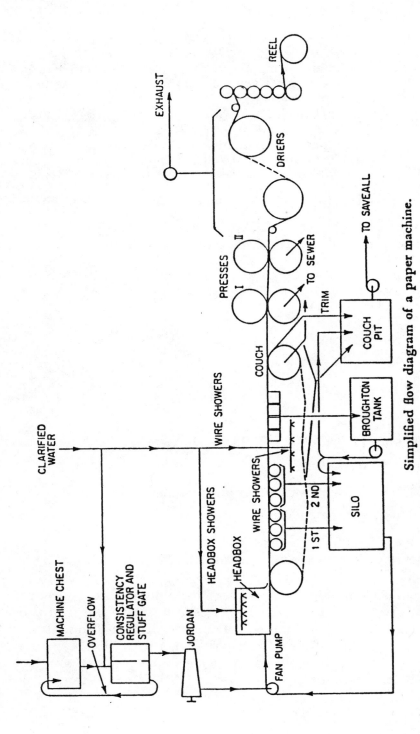

Figure 1.2 Fourdrinier process.

head box, and the material is deposited on this belt. As it is processing toward the calendering rolls, vacuum is used to pull out the water. The material then is compressed between the two presses. At this point, it must have sufficient strength to hold together on the dryers. The material is made in coils.

The following are associated with the beater addition process:

1. Fibers and fillers are added to water in a beater chamber (called a machine chest) at a low concentration, typically less than 5 percent solids.
2. Fibers and fillers are agitated to form a suspension. The latex binder is then added, as well as:
   Pigments (if used)
   Vulcanizing agents
   Aging retardants, antioxidants
   Disposition aids, etc.
3. Along the way to the head box, the suspension is usually diluted to 1 to 2 percent solids.
4. The head box evenly distributes the suspension onto the moving fourdrinier wire screen.
5. As the wire screen travels, water is drained from the suspension by gravity and suction boxes, leaving the binder, fiber, and fillers deposited on the screen.
6. By the time the paper "slurry" has reached the end of the machine wet end, enough water has been removed so that the raw paper becomes self-supporting and can be stripped free from the screen.
7. The paper is then passed through a series of steam-heated "drying cans" to remove all but 1 to 2 percent residual moisture. Infrared or additional heat sources may be used to assist the drying process.
8. The paper is then passed through the calender stack, to compress the material to the proper gauge and density.

Figure 1.3 depicts a similar process. This is called the cylinder process where there is a vat containing the raw materials at the forming board. The material is picked up on the cylinder and is deposited continuously until the proper thickness is achieved. The materials made in the cylinder process can be considerably thicker than those made on the fourdrinier.

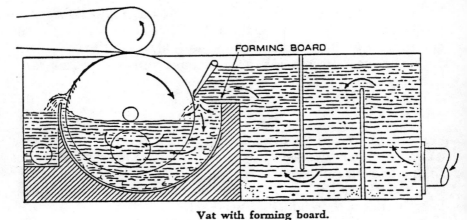

Figure 1.3 Cylinder process.

Compressed process. The compressed material process is shown schematically in Figure 1.4. Here there are two calendering rolls of different diameters. The smaller roll is kept cold while the larger roll is heated. The dough containing the material's ingredients, composed of rubber and binders, is placed between them. The calendering rolls squeeze the ingredients. The material deposits on the larger roll owing to its higher

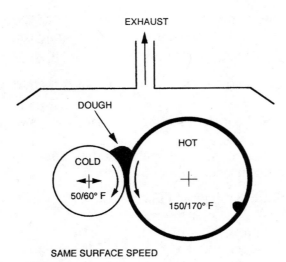

Figure 1.4 Compressed material process.

temperature. As the material thickens, the smaller roll moves outward. Upon reaching the proper thickness, a skiver is used to cut the sheet and pull it off the large roll. This is a solvent process and therefore must be equipped with exhaust systems.

The following is associated with the calendering process:

1. Rubber binder is masticated to a pastelike consistency with an organic solvent.

2. Binder is chosen based on the intended use, e.g., neoprene or nitrile for oil or solvent applications, SBR for high swell in oil application, polyisobutylene for acid resistance.

3. As the rubber is being mixed, other ingredients are incorporated. These include:
   Fibrous reinforcing materials
   Fillers
   Pigments (if used)
   Vulcanizing agents
   Vulcanizing adjuvants such as ZnOMgO
   Aging retardants, antioxidants
   Deposition aids, etc.

The rubber binder generally comprises 10 to 15 percent of the product, and fibers and fillers make up most of the balance.

4. The mixture is dumped between two large cylindrical rollers of different diameters moving at different speeds relative to their surfaces. The dough begins to build up on the larger, heated roller. The smaller roller is chilled to avoid accumulating any dough buildup.

5. The smaller roller regulates the gauge of the sheet being built.

6. A movable feed roll is used to regulate the distance between the large and small rollers.

7. Most of the organic solvent is eliminated while the product is on the roller. Some vulcanization can also take place.

8. When the desired thickness is obtained, the rollers are stopped and the sheet is stripped off.

9. The sheet is usually dried to remove residual solvent. An antistick coating is often applied to the sheet.

Graphite process. The fourth material-making process is associated with what is called expanded or flexible graphite. The mined graphite is acid treated and then heated to a very high temperature. This results in exfoliated worms, which are later placed on a web and compressed in a calendering line. This results in rolls of the graphite material. There is no binder, as the material adhesively bonds to itself. Figure 1.5 depicts this process.

Reinforced process. The fifth process used to make gasket materials utilizes a reinforced support sheet or plate along with a process similar to the beater addition of the fourdrinier process. In this case, the material is placed on the support sheet or plate, usually cold-rolled steel, and is calendered and combined to it. After heating, which is done for curing purposes, the material is coiled. Figure 1.6 depicts this process.

Nonasbestos Gasketing Materials

For many years, asbestos was used for a wide variety of gasketing materials. However, in the last decade and a half, owing to environmental and health considerations, asbestos has been virtually eliminated from most gasket materials. The typical asbestos formulation had approximately 80 percent asbestos fibers, 5 to 15 percent of rubber binders, and a small percentage of other fillers, miscellaneous chemicals, and processing agent ingredients.

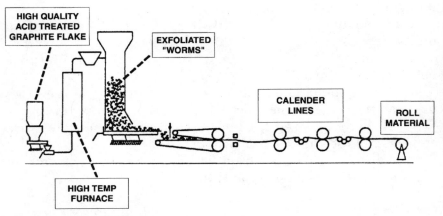

Figure 1.5 Expanded graphite process.

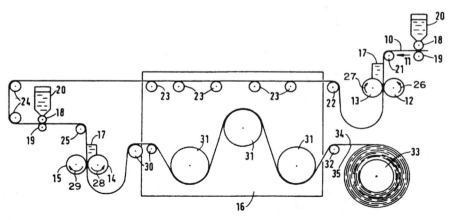

Figure 1.6 Nonstick nonasbestos perforated core process.

A wide variety of nonasbestos formulations have been developed utilizing an even wider variety of ingredients. Generally, they contain about the same amount of the rubber binder, but they have anywhere from 5 to 20 percent of some fiber's pulp and fillers such as inorganic clays which can be up to 70 percent of the formulation. As with asbestos, there are also a number of small amounts of miscellaneous ingredients for processing antioxidants, etc.

The functions of the various ingredients are as follows:

Fibers:

- Heat resistance
- Fluid resistance
- Tensile strength

Rubber binders:

- Bonding of ingredients
- Fluid response (resistance and/or swell)
- Recovery

Fillers:

- Crush and extrusion resistance
- Heat resistance
- Material void structure

Miscellaneous ingredients:
- Heat resistance
- Fluid resistance
- Crush and extrusion
- Cross linking of the ingredients
- Die cutting improvements

As noted, a wide variety of nonasbestos gasketing materials are in the marketplace today. Because of this, close communication between gasket designers and gasket suppliers is required.

In the beater addition process, a slurry of water, rubber latex, fibers, and fillers is deposited on a belt, the water is drawn off, and a gasket material is made. After the water is drawn off, there are voids in the material. These voids must be closed or filled for the gasket to seal. The clamp load of the application closes the voids. Some enhancements added to gaskets to fill these voids are identified later.

In the calendering process, uncured rubber, solvents, fibers, and fillers are squeezed together between the calendering rolls. The compound attaches to one of the rolls and is removed when the proper thickness of the material is achieved. Sheet materials made by the process are called compressed products. They do not contain the void content of the beater sheets but many times still need sealing enhancements owing to poor clamping load distribution and the low compressibility inherent with these materials.

The graphite material contains no binders and therefore does not possess high tensile strength. Careful handling of the material is required. Its resistance to oil and coolant pressure is not high, and aids to improve the fluid resistance to these fluids are therefore many times utilized.

The reinforced material is generally of very high density and, like the compressed materials, may require sealing aids owing to the low compressibility associated with the high density.

Important Material Characteristics for Processing and/or Assembly

A number of sheet material characteristics are important to the fabricator for processing sheets into gaskets and/or for the end user who installs the gasket on the joint. These characteristics follow.

Dust and sliver amounts

During blanking of the sheet material, a lack of dust is desired, as it causes many problems during processing. Dust gets into bearings and other machine components, resulting in premature repair, downtime, and reduction of productivity. In addition, dust on the gasket results in adherence difficulties of subsequent operations such as coating and printing. Since the conversion to nonasbestos, dust has become a major problem with gasket materials with the incorporation of inorganics. Dust is also a major complaint of workers.

Slivers are another characteristic of nonasbestos materials. Slivers are little pieces of the material that occur during blanking. They can cause dents in the gasket, resulting in potential leak paths. Sliver removal is costly, and sliver-free material is therefore highly desired by gasket fabricators.

Tool wear and life

The conversion of the asbestos-based sheets to the nonasbestos-based sheets has resulted in significant tool wear increases. Reports of decreased tool life abound. Some fabricators indicate a reduction of tool life by factors of $\frac{1}{3}$ to $\frac{1}{20}$. The exact reason or reasons for the increased wear are not known, but many gasket manufacturers are conducting investigations of tool wear. The Gasket Fabricators Association (GFA) also is conducting an investigation in die wear.

Scuff resistance

The sheet material's ability to resist scuffing is important to the gasket fabricator. The rough handling and transporting of materials during fabrication can result in scuffing of surfaces of the gaskets. Scuffing can cause leakage paths, thus rendering sealing performance questionable.

Breaking strengths

The breaking strength of the material must be sufficient to resist fracture during processing. Some of the processing operations exert considerable tensile pull on the material. Breaking strengths of compressed materials are much higher than those of the beater addition and graphite materials.

Handling characteristics

The handling characteristics of the gasket material are important during processing, and these same characteristics are also important during assembly of the gasket. Rigidity of the gasket, for example, may be important for ease of assembly and insurance of proper installation. In some cases, the gasket may be installed by robots, and this type of handling has to be taken into account in gasket material selection.

Material Properties That Are Important in Gasketing

The following are properties of the gasket material that are important for sealing performance in the application:

- Chemical compatibility—to be resistant to the media being sealed.
- Heat resistance—to withstand the temperature of the environment.
- Sealability—to provide sealing ability both through the material and over its surface.
- Compressibility or macroconformability—to conform to the distortions and undulations of the mating flanges.
- Microconformability—to "flow into" the irregularities of the mating flanges' surface finishes.
- Recovery—to follow the motions of the flanges caused by thermal or mechanical forces.
- Creep relaxation—to retain sufficient stress for continued sealing over an extended period of time.
- Erosion resistance—to accommodate fluid impingement in cases where the gasket is required to act as a metering device.

- Compressive strength—to resist crush and/or extrusion caused by high stresses.
- Tensile or radial strength—to resist blowout due to the pressure of the media.
- Shear strength—to handle the shear motion of the mating flanges due to thermal and mechanical effects of the mating flanges.
- Z strength—to result in easy used gasket removal without internal fracture of the material.
- Flexibility—to be able to flex without fracture both initially and after extended storage.
- Antistick—to ensure gasket removal without sticking.
- Heat conductivity—to permit the desired heat transfer of the application.
- Acoustic isolation—to provide the required noise isolation of the application.
- Anticorrosiveness—to provide anticorrosive characteristics between the material and the mating flanges.
- Dimensional stability—to permit correct assembly.

To ensure that there will be no health problems for the fabricator and/or end user, the material must be nontoxic. This is a most important characteristic of the material.

Another important characteristic of the gasket material, as noted earlier, is sealability. This, however, is not only dependent upon the gasket's material properties but also is a function of the environmental conditions such as clamp load, bolt span, and flange rigidity. Environmental conditions are discussed later.

Another sheet material characteristic that is not necessarily application-important but is required for general sealing performance is uniformity of the material in regard to consistency of formulation thickness, density, and surface finish. Some of the reasons for this uniformity are:

- Consistency of formulation is vital for compatibility of the material with the medium being sealed, as well as for other reasons.

- Many gaskets have precise requirements, as they are shims in addition to being gaskets. Thickness tolerance and density uniformity from lot to lot are important for control of compressed thickness control.
- Uniformity of density is also very critical for some of the fabricators' added enhancements such as saturating, laminating, and surface coating.
- Surface finish consistency within a lot and from lot to lot is important for consistent surface sealing of the mating flanges.

The primary criterion for a material to be impervious to a fluid is to achieve a sufficient density to eliminate voids which might allow capillary flow of the fluid through it. This requirement may be met in two ways: by compressing the material to fill the voids and/or partially or completely filling them during fabrication by means of binders and fillers. Also, to maintain its impermeability for a prolonged time, the constituents of the material must be able to resist degradation and disintegration resulting from chemical attack and/or temperature of the application.

Most gasket materials are composed of a fibrous or granular base material, forming a basic matrix or foundation, which is held together or strengthened with a binder. The choice of combinations of binder and base material depends on the compatibility of the components, conditions of the sealing environment, and load-bearing properties required for the application.

The rubber binder forms the matrix which holds together all the components that comprise compressed sheet and beater-addition sheet packing. The type of rubber used will determine, in large part, the behavior and performance of the gasket material when exposed to automotive and diesel engine fluids. The binder also influences heat resistance, sealability, compressive strength, extrusion resistance, and creep relaxation.

Binder coverage of the reinforcing fibers affects properties of the gasket material. Binder used for compressed sheet is blended with solvent. When mixed in, the fibers get completely coated with a layer of rubber. This yields a sheet with low porosity and good flexibility and handling properties. However, the rubber-to-rubber contact between fibers may lessen resistance to extrusion; the rubber can act as a "lubricant" under high-load conditions. Beater-addition products are made using a latex emulsion. The rubber is deposited as discrete particles onto the fibers; the product retains direct fiber-to-fiber contact, improving extrusion resistance.

Fiber Materials

The following are the functions of the various gasket material ingredients:

Fibers

- Provide tensile strength in the x, y, and z dimensions
- Provide media resistance
- Provide heat resistance
- Aid in processing

Fillers

- Provide the necessary void arrangement for desired sealability and/or saturation
- Provide crush resistance
- Provide heat resistance

Binders

- Bond the ingredients of the material
- Provide desired fluid resistance and/or control
- Provide resilience

Other additives

- Aid in cutting
- Aid in cross linking of the binder
- Provide fluid resistance
- Provide heat resistance

Various types of fibers are used. The most prevalent are cellulose and polyamide. Others such as acrylics, polyimides, carbons, and other inorganic fibers are also utilized. Asbestos was a popular fiber in the past but is almost completely eliminated from gaskets today.

Fibers are the backbone of the gasket material. They control its compressive strength and extrusion resistance. Fibers coupled with the binder affect the relaxation and heat resistance of the material.

Cellulose. Cellulose is a nonasbestos fiber that has good chemical resistance to most fluids except strong acids and bases. The temperature limitation is approximately 300°F. Changes in humidity may result in dimensional changes and/or hardening.

Asbestos. This material has good resistance to 800°F and is noncombustible. It is almost chemically inert (crocidolite fibers, commonly known as blue asbestos, resist even inorganic acids) and has very low compressibility. The binder dictates resistance to temperature and medium to be sealed. This fiber is permitted for use only in special industrial applications where safety of the gasketed joint is involved and where no other fiber can reliably seal. As mentioned earlier, asbestos fibers have essentially been eliminated from gasketing materials.

Nonasbestos. A number of other nonasbestos fibers are being used in gaskets. Some of these are glass, carbon, polyaramids, acrylics, ceramics, and various inorganic fibers. Temperature limits from 750 to 2400°F are obtainable. Use of these fillers is an emerging field today, and suppliers should be contacted before these fibers are specified for use. The nonasbestos formulations differ considerably from the asbestos formulations that they have replaced.

A typical formulation of an asbestos product consisted of 70 to 90 percent of asbestos fibers, 10 to 20 percent of a latex binder, and a few other ingredients such as processing agents, antioxidants, plasticizers, and fillers. The binders were generally nitrile-butadiene (NBR) or styrene-butadiene (SBR); blends of them were also utilized.

The nonasbestos formulations significantly differ from the typical asbestos formulations. They contain 5 to 20 percent fibers, about the same percentage of binder, 10 to 20 percent, with the remainder being fillers along with some processing ingredients as noted above.

The depositing of the binder onto any fiber requires a positively charged surface of either the fiber or the filler. Most fillers have a negative charge; thus the surface must be neutralized and then positively charged. The fibers are used to provide tensile strength, heat resistance, and fluid resistance and to aid in processing. The binders are used to "bond" the components. This determines the material's reaction to fluids as well as its resiliency. The fillers are used to provide the desired void structure and to impart crush extrusion resistance as well as heat resistance. The other ingredients are also used for crush extrusion resistance as well as resistance to the fluid. Some are used to cross-link the binder and others are used to improve die cutting.

Binders

In addition to natural rubber, a wide variety of synthetic rubbers are used for gaskets. Some of the most common types that are utilized are:

Natural—NR

Nitrile (Buna N)—NBR

Styrene butadiene (Buna S)—SBR

Polychloroprene (neoprene)—CR

Isobutylene (butyl)—BR or IIR

Silicone—MQ, VMQ, and SI

Fluorosilicone—FSi

Chlorosulfonated polyethylene (Hypalon)—CSM

Ethylene propylene—EPM, EPDM, and EPR

Fluorocarbon (Viton, Technoflon, Fluorel)—FKM

Polyurethane—PU, AU, and EU

Polyacrylic (Hycar, Krynac, Thiacril, Cyanacryl)—ACM

These binders provide the bonding of the ingredients, and the heat and fluid resistance of the material. Some of the properties of these binders are:

Natural. Good mechanical properties and is impervious to water and air. It has uncontrolled swell in petroleum oil and fuel and chlorinated solvents. The temperature range is −60 to 250°F.

Nitrile. Excellent resistance to oils and dilute acids. It has good compression set characteristics and has a temperature range of −60 to 275°F.

Styrene. Similar to natural rubber but has slightly improved properties. Not recommended for petroleum oils owing to its high swell when exposed to them. Also, it is not recommended for exposure to sunlight. The temperature range is −60 to 250°F.

Neoprene. Good resistance to water, alkalies, nonaromatic oils, and solvents. Its temperature range is −60 to 275°F.

Butyl. Excellent resistance to air and water, fair resistance to dilute acids, and poor resistance to oils and solvents. It has a temperature range of −60 to 225°F.

Silicone. Good heat stability and low-temperature flexibility. It is not suitable for high mechanical pressure. It is not recommended for ketone solvents or hydrocarbon fuels. Its temperature range is −100 to 500°F.

Fluorosilicone. Resistant to hydrocarbon fuels and oils. Good in fuel systems. Temperature range is −100 to 350°F.

Chlorosulfonated polyethylene. Excellent resistance and also has good resistance to flame. Its temperature range is −60 to 225°F.

Ethylene propylene rubber. Excellent resistance to hot air, water, coolants, and most dilute acids and bases. It swells in petroleum fuels and oils without severe degradation. The temperature range is −60 to 300°F.

Fluorocarbon. Good resistance to oils, fuel, and chlorinated solvents. Poor resistance to alcohols, ketones, and acids. Its temperature range is −20 to 450°F.

Polyurethane. Excellent resistance to oils, solvents, and ozone. It has a limited temperature range of −60 to 225°F.

Polyacrylic. Excellent resistance to mineral oils, hypoid oils, and greases. The temperature range is 0 to 350°F.

Some resins are also used as binders. They usually possess better chemical resistance than rubber. Temperature limitations depend on whether the resin is thermosetting or thermoplastic.

Filler Materials

In some cases, inert fillers are added to the material composition to aid in filling voids. Some examples are barytes, clay, mica, graphite, glass beads, and silicates. In addition to influencing the properties of the material, they can reduce cost.

Nonfiber Gasket Materials

A number of gasket materials are not of the fiber category. Some of the more popular are:

Rubber. Rubber materials provide varying temperature and chemical resistance depending on the type of rubber used. These rubber and rubberlike materials are used not only as binders but in some cases as complete gaskets. Nitrile, neoprene, polyacrylic, and silicone are the most popular rubbers used as gasket materials. Temperature limitations are the same as noted earlier under Binders.

The gaskets can be cut from cured extruded or molded sheets. In many cases gaskets are molded out of these materials. These products are discussed later.

Cork and cork-rubber. High compressibility allows easy density increase of the material, thus enabling an effective seal at low flange pressures. The temperature limit is approximately 250°F for cork and 300°F for cork-rubber compositions. Chemical resistance to water, oil, and solvents is good, but resistance to inorganic acids, alkalies, and oxidizing environments is poor. It conforms well to distorted flanges.

Tanned glue and glycerin. This combination produces a continuous gel structure throughout the material, thus allowing sealing at low flange loading. It has good chemical resistance to most oils, fuels, and solvents. It swells in water but is not soluble. The temperature limit is 200°F. It is used as a saturant in cellulose paper.

Plastics. This is a newer variety of gasket materials which generally resist temperatures and corrosive environments better than rubbers. Two of the popular materials are:

Teflon. Polytetrafluoroethylene—Teflon—was developed by E. I. du Pont de Nemours.

Kel-F. Trifluorochloroethylene—Kel-F—was developed by M. W. Kellogg Co.

Metallic Gaskets

Many metals are used for gasketing purposes. Some of the most common range from soft varieties such as lead, copper, steel, nickel, stainless, and inconel to the high alloyed steels. Noble metals such as platinum, silver, and gold also have been used to a limited extent.

Metallic gaskets are available in a wide array of designs that are both standard and custom designs. Some are to be used unconfined while others are used in a confined position. Both elastic and plastic sealing is utilized. Some metallic designs use the internal pressure to improve the sealing.

There are plain rings which fit in grooves, as well as self-energized rings incorporating holes or slots so that the internal pressure activates the seal, and also rings which are pressure-filled with inert gases. Figure 1.7 depicts some types of semi-metallic and all-metallic gaskets.

With a few exceptions, metallic gaskets take a permanent set when compressed in an assembly, and, unlike nonmetallic gaskets, have little or no recovery to compensate for contact face separation. Therefore, joints in which metal gaskets are to be used must have sufficient rigidity to assure a minimum of bending both during the initial bolting up and when the assembly is operating.

Each type of metallic gasket will perform best when the flange contact faces have a specific surface finish. Depending upon gasket type, this optimum surface finish requirement

Gasket Types

Name	Type	Material
CORRUGATED		Aluminum Copper Soft Steel (Iron) Monel Stainless Steel
CORRUGATED AND COATED		Aluminum Copper Soft Steel (Iron) Monel Stainless Steel
SPIRAL WOUND	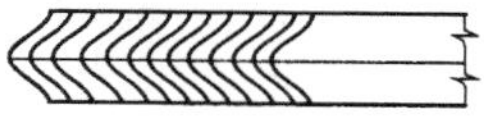	Carbon Steel Carbon Steel Stainless Steel Stainless Steel
JACKETED	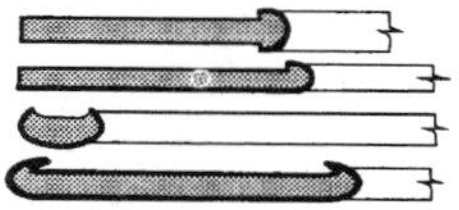	Lead Aluminum Copper Soft Steel (Iron) Nickel Monel Stainless Steel
JACKETED AND CORRUGATED		Lead Aluminum Copper Soft Steel (Iron) Monel Stainless Steel
FLAT METAL		Aluminum Copper Soft Steel (Iron) Monel Stainless Steel
MACHINED METAL		Aluminum Copper Soft Steel (Iron) Monel Stainless Steel
METALWIRE	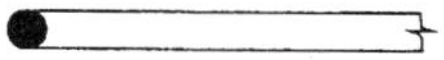	Aluminum Copper Soft Steel (Iron) Stainless Steel
ENCLOSED WIRE	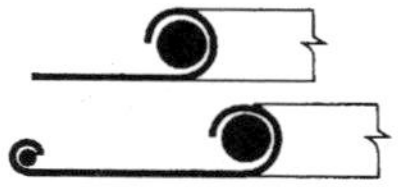	Aluminum Jacket Aluminum Cores Aluminum Jacket Stainless Steel Cores Stainless Steel Jacket Stainless Steel Cores

Figure 1.7 Various semimetallic and all-metallic gaskets.

may be very smooth or deeply tooled. When a minimum gasket seating stress is given for a particular type of gasket, it applies only when the flanges have the optimum surface finish. Any deviation from this optimum finish will increase the seating stress required.

In high-pressure, clamp load, and temperature applications, a high-spring-rate (high stress per unit compression) material is necessary to achieve high loading at low compression in order to seal the high pressures involved. These applications generally rely on sealing resulting from localized yielding under the unit loading. In addition to the high spring rate, high heat resistance is mandatory. To economically satisfy these conditions, metallic gaskets are most commonly used.

In applications where close tolerances in machining (surface finish and parallelism) are obtainable, a solid steel construction may be used. In those situations where close machining and assembly is not economical, it is necessary to sacrifice some gasket rigidity to allow for conformability. In such cases, conformability exceeding that resulting from localized yielding must be inherent in the design. The material can be embossed or coated to provide the required conformity.

The maximum service temperatures of various platings or coatings used with metal gaskets are as follows:

Coating	Maximum recommended service temperature, °F
Indium	275
Cadmium	400
PTFE	575
Silver	1475
Gold	1500

Uncoated rings usually are suitable for liquid sealing applications. Coated rings are sometimes needed for gas sealing cases.

Since such a wide variety of designs, materials, and platings are used in metallic gaskets, it is recommended that the reader directly contact metallic gasket suppliers for design and sealing information.

The maximum service temperatures of various metallic gaskets are as follows:

Material	Maximum recommended temperature, °F
Lead	200
Brass	500
Copper	600
Aluminum	800
Steel	1000
Nickel	1400
Stainless steel:	
304	800
316	800
410	1200
309	1600
321	1600
347	1600
Monel	1500
Inconel	2000

Gasket Types

Corrugated. Consists of thin metal, corrugated or with embossed concentric rings. They are used plain, coated with gasket compound, or with nonmetallic fillers cemented in the corrugations. This type of gasket requires the least costly tooling for nonstandard sizes or irregular shapes. By proper material selection, they can be used at practically any temperature.

Corrugated gaskets are essentially a line-contact seal. Multiple concentric corrugations provide a labyrinth effect, along with mechanical support for the gasket compound or asbestos-cord inserts if they are used. Corrugations provide some degree of resilience, depending on their pitch and depth and the type and thickness of metal used.

Corrugations. Although a minimum of three concentric corrugations is desirable on each gasket face, many applications exist using only one. A slight flat inside the inner corrugation and beyond the outside corrugation will stiffen the gasket. This form is desirable if space is available. For full-face gaskets in thin, lightly bolted flanges, one to three corrugations inside of the bolt circle and one or more outside of the bolt circle will equalize compressive stresses and may be helpful in preventing flange distortion.

Jacketed, soft filler. Consists of soft compressible filler partially or wholly encased in a metal jacket. These gaskets are more compressible than corrugated types. They offer better compensation for flange irregularities when higher pressures are to be sealed.

The primary seal against leakage is the inner metal lap, where the gasket is thickest when compressed. This area cold-flows, effecting the seal. The entire inner lap must be under compression. The outer lap, if any, provides a secondary seal between flange faces when compressed. Intermediate corrugations may contribute to the labyrinth effect.

These gaskets are used for noncircular as well as circular applications. This gasket type is used for applications at temperatures up to those which limit the filler and metal endurance. Because of limited resilience, these gaskets should be used only in assemblies in which the elasticity of the bolts or other factors can compensate for joint relaxation. They should not be used in joints requiring close maintenance of the compressed thickness.

Jacketed gaskets have certain limitations as far as lap width is concerned. The maximum lap width is that which can be drawn and folded over without cracking or wrinkling. This is a function of the metal, its thickness, the gasket thickness, and the gasket diameter. The maximum lap width must be taken into consideration when relatively small diameters or radii are to be formed.

Spiral-wound gaskets. These consist of V-shaped, preformed plies of metal, wound up in a spiral with soft nonmetallic fillers such as Teflon, aramid fiber, or graphite. The V shape gives unique, springlike characteristics.

These gaskets have good recovery. Density of construction can be controlled to provide an optimum seal under various bolting conditions. They have good tolerance for flange–surface finish irregularities and are furnished in a wide variety of metals in circular and limited noncircular shapes.

Sealing action is the result of a combination of the flow of the metal and soft filler plies when the gasket is compressed. Inner and outer metal-to-metal plies must be under compression.

Spiral-wound gaskets are used for any circular or moderately noncircular application. They are particularly suited for assemblies subject to extremes in joint relaxation, temperature or pressure cycling, shock, or vibration. Excellent performance in joints restricting compression is provided, since their resiliency will compensate for modest flange separation.

The spiral-wound gasket gives best performance when compressed to a predetermined specific thickness. Its compressibility can be accurately controlled for a specific bolt loading by varying the density, i.e., the number of metal asbestos plies or wraps per unit of gasket width.

Spiral-wound gaskets are made in most metals. The performance of a spiral-wound gasket depends upon the springlike action of the V-shaped metal strip. Therefore, the metal used should be one that will best maintain its resilience at the operating temperature.

The maximum width of a spiral-wound gasket is a function of the diameter and thickness. In general, the larger the diameter is, the narrower the gasket is. The spiral-wound gasket requires more careful dimensioning in relation to flange facing to assure that the inner and outer layers of metal plies are under compression between the flange facings.

Spiral-wound gaskets can be furnished in moderately noncircular shapes depending upon size and desired configuration. As a very rough guide for shapes approximating an oval, the major axis should not exceed twice the minor axis. Straight-sided rectangles (radiused corners) with an inside measurement in excess of about 6 in may not be practical to fabricate.

Although spiral-wound gaskets can be used for general, noncritical service with almost any commercially produced flange surface finish, 150 to 200 rms is preferred. For critical service use a finish of 32 to 80 rms.

Flat metal gaskets. These are defined as gaskets that are relatively thin compared to their width. They can be used as cut from sheet metal or with the surface area reduced by machining to improve sealability.

Plain metal, washer-shaped gaskets are relatively inexpensive to produce and can perform satisfactorily in a variety of applications over a wide temperature range. The machined types, with reduced surface area, may be the answer to high-pressure, high-temperature, or highly corrosive applications in nonspecial flanges where bolting forces are too light for the plain flat type.

All types seal by flow of the gasket surface caused by brute-force compressive loads. Loads actually must exceed the tensile strength of the gasket metal on the gasket-contact area. Therefore, surface finish of both flange and gasket is very important.

Plain metal gaskets are normally furnished as cut from the metal in the "as received" condition. Therefore, the gasket will have the surface finishes caused by mill rolling, plus any storage or in-transit damage. Furthermore, depending on the method of manufacture, the gasket edges may have burrs or other irregularities which may or may not affect performance. If higher quality levels are required, such as no burrs, scratch-free surfaces, or close decimal tolerances, cost is increased.

For best performance, flat plain-metal gaskets should be used between flange faces with concentric serrated surfaces. If this is not practical, a very light cut spiral tool finish of 80 rms may be used.

Machined metal. These are usually rectangular- or triangular-shaped gaskets, machined cross sections from solid metal. They are used for high-pressure and high-temperature services where operating conditions require a special joint design.

These gaskets usually seal by virtue of a line contact or wedging action which causes surface flow. Some of them are pressure-actuated; i.e., the higher the pressure, the tighter the joint.

Metal wire. These gaskets are generally made from round wire of the desired diameter cut to the length of the gasket mean circumference, then formed into a circle and welded. They provide positive pressures. Since only line contact occurs, they have high local seating stress at low bolt loads. The contact faces increase in flowing into flange faces.

Round, solid gaskets are essentially used on equipment designed specifically for them. Flanges are usually grooved or otherwise faced to accurately locate the gasket during assembly. However, in some applications they are used between flat faces.

Any wrought or forged material is usually suitable for this type of gasket. Cast material should not be considered. Materials requiring heat treatment after final machining usually should not be used because of possible warpage.

References

Armstrong Gasket Design Manual, Armstrong Cork Co., Lancaster, Pa., 1978.
Buchter, H. Hugo: *Industrial Sealing Technology,* Wiley, New York, 1979.
Passarella, M. T.: "Designing with Thermal Insulating Gasketing and Gasketing Systems," SAE paper 830219, presented at the SAE International Congress and Exposition, February 1983.
Rogers, R., R. Foster, and K. Wastler: "Factors Affecting the Formulation of Nonasbestos Gasket Materials," SAE paper 910206, February 1991.
Standard Classification System for Nonmetallic Gasket Materials, ANSI/ASTM F104-70a, American Society for Testing and Materials.
Testing Gasket Performance, Centre de Developpement Technologique, Ecole Polytechnique de Montreal.

Chapter 2

The Gasket and Its Environment

Gasket Design and Environmental Conditions

Gasket design

As stated previously, gaskets are used to create and maintain a seal between two separable flanges. In theory, if the flanges were perfectly smooth, parallel, and rigid, they could be bolted together and would seal without a gasket. But in practice, flanges have rough surface finishes and limited rigidity. They are not perfectly parallel and may be secured by bolts of varying lengths and not uniformly spaced around the flanges. Reflecting these conditions, flange loading is often very nonuniform. The gasket must compensate for this nonuniform flange loading and distortion. It also must conform to the flange surface irregularities.

Once it is realized that the insertion of a gasket between two flanges is necessary, a host of design problems must be met. The first is the recognition that the medium being sealed may be corrosive to the gasket material. In addition, the pressure of the medium being sealed exerts radial forces on the gasket, tending to blow it out. This pressure can also exert forces on the assembly, tending to open the flanges and reduce the sealing stress on the gasket. Furthermore, the gasket and its environment are likely to be subjected to large variations in temperature, and thermal distortions ultimately will therefore occur. Finally, under the influence of the sealed medium, the operating temperature, and the pressure the gasketing material may change dimension, because of its creep-relaxation properties, and lower bolt torque and sealing stress.

The basic factor in the creation of the seal is sufficient stress on the gasket to ensure its conformation to the flange surfaces. This blocks the passage of the media between the gasket and the flange. In addition, this stress must be high enough to close any voids in the basic material if it is to block the passage of sealed media.

Some of the factors considered when creating the seal are:

1. Selection of gasket material with regard to its compressibility.
2. Determination of the pressure and temperature of the assembly.
3. Flange flatness, surface finish, and rigidity.
4. Gasket configuration, especially in regard to bolt locations. After the seal has been created, it must be maintained. Some relevant factors follow.

A gasket most often is a viscoelastic material. It may be represented by the mechanical model shown in Fig. 2.1, comprised of a spring in series with a parallel combination of a spring and dashpot which in turn is in series with a dashpot.

This model then may be used to explain the gasket's load-deflection characteristics. A viscoelastic gasket is loaded with a force P lb for a certain time. Compression of the elements A, B, and C of the model occurs. Referring to Fig. 2.2, the first spring (A) is known as the elastic response. This deflection is immediately recoverable when the force P is removed. The second element (B), the series-parallel combination, is known as the viscoelastic response. This deflection is time-dependent but is totally recoverable. When the force P is removed, this element will return to its original configuration after some period of time. The third element (C) is known as the viscous response; its deflection is time-dependent and totally unrecoverable. Changes in this element account for compression set or creep relaxation.

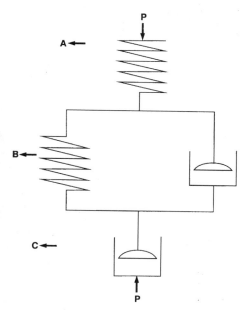

Figure 2.1 Mechanical model of a typical gasket.

The gasket, owing to its viscoelastic properties, will change its load-deformation properties with temperature. Compression set is a common source of fibrous and rubber gasketing problems. When a gasket undergoes permanent relaxation, sealing stress on the gasket is lowered. If this occurs in a poorly loaded joint or in a flange where there is nonuniform bolt spacing and an inherently poor bolt loading pattern, the sealing stress may well fall below that minimum stress required to seal the medium. The joint will then leak.

Minimum seating stresses for various gasket materials should be obtained from the material fabricator or gasket supplier. In addition, the load remaining on the gasket during operation must remain high enough to prevent blowout of the gasket. During operation, the hydrostatic end force, which is associated with the internal pressure, tends to unload the gasket and could result in leakage or blowout. Maintaining of the seal during operation is discussed later.

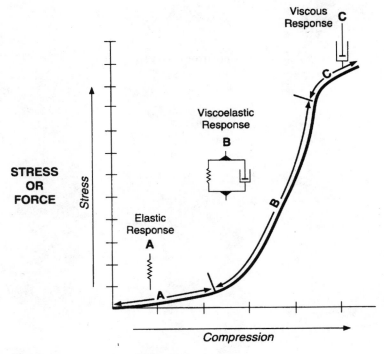

Figure 2.2 Gasket force deflection.

Environmental conditions

Many environmental conditions and factors influence the sealing performance of gaskets. Flange design details, in particular, are most important. Design details such as number, size, length, and spacing of clamping bolts; and flange properties such as thickness, modulus, surface finish, waviness, and flatness are important factors. In particular, flange bowing is a most common type of problem associated with the sealing of a gasketed joint.

Different gasket materials and types require different surface finishes for optimum sealing. Soft gaskets such as rubber can seal very rough surface finishes in the vicinity of 500 μin. Some metallic gaskets may require finishes as fine as 16 to 32 μin for best sealing. Most gaskets, however, will seal in the surface finish range of 60 to 125 μin. There are two main reasons for the surface finish differences:

1. The gasket must be able to conform to the roughness for surface sealing.
2. It must have adequate conformability into the mating flange to create frictional forces and thereby resist radial motion due to the internal pressure. This is necessary to prevent blowout.

In addition, elimination of the radial motion will result in maintaining the initial clamp-up sealing condition. The radial motion is micro in amount but can result in localized fretting, and a leakage path may be created.

Some trial-and-error gasket designs for specific applications result from gasket sealing complexity due to the wide variety of environmental conditions. Information on gasket design and selection is included later. This information will enable a designer to minimize the chance for leaks. Since the gasketed joint is so complex, adherence to the procedure will not ensure adequate performance in all cases. If and when inadequate gasket performance occurs, gasket manufacturers should be contacted for assistance.

The function of the fastener in a gasketed joint is to apply and maintain the load required to seal the joint.

Threaded nuts, bolts, and studs are the usual fasteners used in industrial gasketing to assemble mating parts. The fastener device must be able to produce a spring load on the gasket to compress it to its proper thickness and density for sealability. The fastener must also be able to maintain proper tension to maintain this compression of the gasket material throughout the life of the assembly. When flanges are made of dissimilar metals, bolting plays a most important part in obtaining and maintaining a satisfactory seal. The fasteners as well as the gasket must be able to compensate for the difference in expansion and contraction of the different flange materials.

Addressing the question of how many bolts or other fasteners can be used involves space available, economic limitations, and flange flexibility considerations as well as getting the required initial load. Approximately 80 percent of the load applied by the fastener may be distributed out along the flange to the midpoint between the bolts. Exactly how this load is distributed is dependent upon the thickness, configuration, and rigidity of the flange. Therefore, it is difficult to provide useful "rule of thumb" guidelines, but cutting the distance between bolts by half will reduce the bowing effect to one-eighth its original value. Conversely, stiffening the flange is frequently more cost-effective than increasing the number of fasteners.

Washers spread the bolting load over a large area and prevent or at least reduce deformation of the flange area directly beneath the head of the bolt. The fit, flatness, and type of washer can have a significant effect upon loading. The relationship of the shank clearance and washer ID chamfer is important, as an increase in surface friction could result in a reduction in bolt loading.

When there is a reduction of clamp load on the gasket, there can be an increase in the movement of the sealing flange components. Therefore, a bolt-locking or thread-locking device may be beneficial.

In some cases star washers or spring washers are used, or one of a variety of liquid or plastic thread-locking compounds. Thread-sealing compounds are also available for applications where the fastener threads into the cavity being sealed. Spring washers can also compensate somewhat for the compression set of the gasket and thus decrease the associated loss of clamp load.

It is essential to avoid overloading the gasket. Gasket materials will crush owing to a combination of compression-, shear-, and extrusion-type displacements of the material. The maximum unit load is a function of the type of material, operating temperature, thickness, and section width, among the principal factors. The designer of the fastener system must know the crush limitations of the gasket material being considered for the application. This information is available from gasket suppliers. Deformation of the flange can also create higher than expected localized loading conditions that can aggravate crush and extrusion.

Flanges must have adequate thickness. Adequate thickness is required to transmit the load created by the bolt to the midpoint between the bolts. It is this midpoint that is the vital point of the design. Maintaining a seal at this location is important and should be kept constantly in mind.

Adequate thickness is also required to minimize the bowing of the flange caused by the bolt loads. If the flange is too thin, the bowing will become excessive and low bolt load will exist at the midpoint. See Figs. 2.3 and 2.4.

Reinforcing flat sheet-metal flanges by adding embosses (sometimes called a tapered stress riser) or by turning up the edges of the flange to create a more rigid beam should be considered when working with thin flanges. See Figs. 2.5 and 2.6.

Another recommendation for a sheet-metal design is to provide a positive backdraft on the flange, concentrating the force

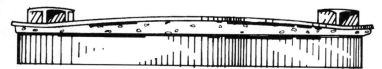

Figure 2.3 Illustration of flange bending or bowing.

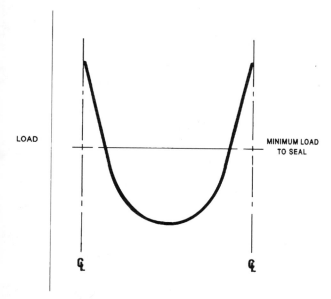

Figure 2.4 Midpoint loading between bolts.

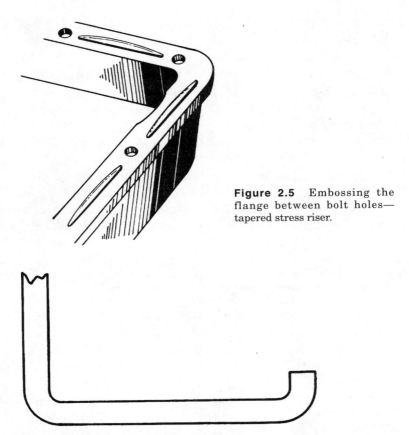

Figure 2.5 Embossing the flange between bolt holes—tapered stress riser.

Figure 2.6 Flange with turned-up edge.

on the inner edge of the gasket. This is shown in Fig. 2.7. It prevents the sealed medium from leaking through the bolt holes, which could happen if the load is concentrated at the outer edge of the gasket.

A flange may also tip or become nonparallel during the clamping process. See Fig. 2.8. The tipping will not give rise to serious performance problems until the gasket compression loads in the area of low compression fall below the minimum load required to seal.

Internal pressure also can create loads on both bolts and flanges to create another type of distortion. For instance, bolts might elongate. This would be elongation in addition to that

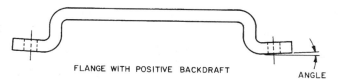

Figure 2.7 Flange with positive backdraft.

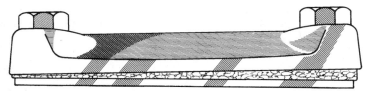

Figure 2.8 Tipping or nonparallelism in a flange.

caused by the initial tightening torques. Yielding of the bolt and unloading of the gasket could result. Also, the flange might deflect or reveal a bowing in addition to the bending caused by the imposition of initial bolt loads. Another force or load created by internal pressure is blowout. Figure 2.9 depicts this blowout as acting on the inner edge of the gasket, tending to push it out from between the flanges.

The number and distribution of the bolts significantly affects the loading pattern between the sealing surfaces of the two flanges. The best clamping pattern is invariably a combination

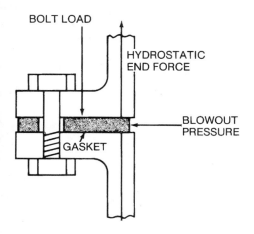

Figure 2.9 Blowout pressure in a gasketed joint.

of the maximum practical number of bolts, even spacing of them, and optimum positioning.

The load-distribution pattern is a series of straight lines drawn from bolt to adjacent bolt. If the sealing area lies on either side of this pattern, it may be a potential leakage spot. The farther the sealing area deviates from this pattern, the more likely the leakage. An example of these various conditions is illustrated in Fig. 2.10.

It is desirable to have an adequate number of bolts, properly spaced, at a joint. Sometimes in certain areas this is not possible. Where the spacing of the bolts must be wide, and larger-diameter bolts cannot be utilized, the designer can help the contours form a uniform width to a sculptured design. Such a revision is shown in Fig. 2.11. Because the width of the gasket is gradually reduced to the midpoint, the unit loading is increased and the midpoint load is better maintained than it would have been if the width of the gasket had not been reduced.

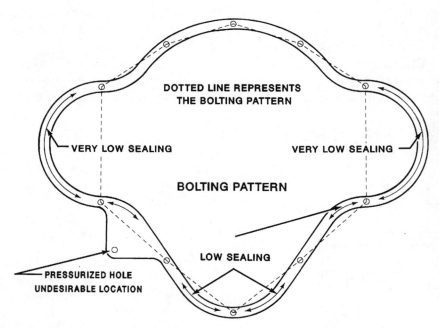

Figure 2.10 Good and poor sealing locations.

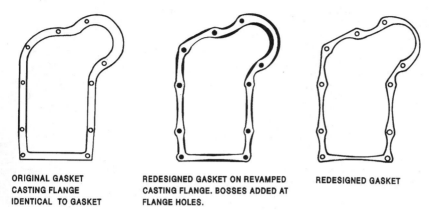

ORIGINAL GASKET
CASTING FLANGE
IDENTICAL TO GASKET

REDESIGNED GASKET ON REVAMPED
CASTING FLANGE. BOSSES ADDED AT
FLANGE HOLES.

REDESIGNED GASKET

Figure 2.11 Sculptured design of gasket—original versus redesigned gasket for improved sealing.

About 80 percent of the bolt load is concentrated around the area of the bolt. Because of the necessity of incorporating a bolt hole in a gasket, the amount of material left at this location may be inadequate to carry the high bolt load, and as a result the gasket may crush. This is especially true with thin sheet-metal flanges where distortion at the bolt holes is common. To alleviate this condition, the flange face and the gasket should be enlarged as shown in Fig. 2.12. This will help compensate for gasket material removed to create the bolt hole.

Very small bolt holes or small noncircular openings create high costs. The centers from such holes will probably require hand picking, and since small holes are easy to miss, extra

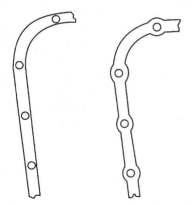

Figure 2.12 Revision to flange and gasket to prevent crush of gasket head bolt holes.

inspection may be needed. It is best to avoid hole sizes under 0.10 in diameter. If the hole is used just for positioning or indexing, a small notch should be considered. See Fig. 2.13.

Ideally, all the bolts of the joint should be torqued simultaneously. Alternately, a good joint can be achieved utilizing proper torque sequences. The sequence in which bolts are tightened has a substantial bearing upon the distribution of the contact area stress. A poorly specified bolt sequence will cause distortion of the flange and result in poor sealing.

The proper bolt-fastening sequence is a crisscross pattern if the bolts are in a circular pattern. See Fig. 2.14.

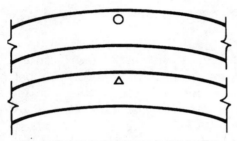

Very small or noncircular holes require extra handling.

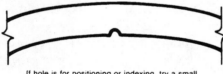

If hole is for positioning or indexing, try a small notch instead.

Figure 2.13 Avoid small and noncircular bolt holes. Consider notching.

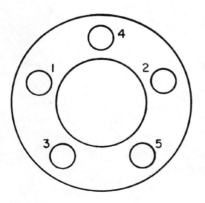

Figure 2.14 Crisscross fastening sequence.

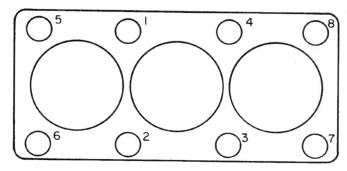

Figure 2.15 Spiral fastening sequence.

If the bolt pattern is noncircular, the fastening sequence is usually a spiral pattern, starting at the middle. See Fig. 2.15.

Tightening of bolts is accomplished by hand or multiheaded automatic torque wrenches. Tightening specifications can be in terms of a specified torque plus additional angle of turn, tightening to yield point of the bolt.

A sealing unbalance can be introduced during assembly. Causes are galled threads or bosses, and unlubricated threads. Either of these conditions can cause low applied bolt tension even though a torque wrench is used. It is well known that lubricated threads are much more efficient in converting torque to bolt tension or clamping force than are dry threads. In some cases, a lubricated bolt can have 50 percent more tension than an unlubricated bolt, even though both were torqued to the same value. The recommended procedure is to oil threads and wipe off thoroughly with a clean, dry cloth. Use hardened washers under bolt heads if bosses are galled; clean up threads on the bolt and retap the threads in the block.

Situations have been known to occur where sufficient metal is removed from warped flanges in resurfacing to cause bolts to bottom in blind holes or to bind on corroded threads below the previous level of tightening. In either case, the gasket is not compressed sufficiently to obtain a complete seal. If excessive machining has been done, this condition should be looked for after tightening of bolts by checking for clearance under bolt heads. If bolt heads are not tight against the boss, the bolts should be removed and washers inserted. In all likelihood a new gasket should be used.

A flange can be resurfaced either too rough or too smooth for best gasket sealing purposes. As noted earlier, good range is between 60 and 125 μin, while 90 to 110 is preferred. Metallic or metal-faced gaskets, although excellent performers, are particularly sensitive to casting finishes either above or below this range. There is low resistance to blowout with very smooth finishes and poor fluid seal with rough finishes.

Surface flatness is also important to good sealing. If flange flatness is excessive, milling or grinding is necessary. Soft surface gaskets are preferred for the extremes since there is greater surface conformity under clamping pressure. All gaskets, whether metal or soft-surfaced, are normally supplied with a coating to aid in sealing rough surface finishes.

In general, gasket manufacturers do not recommend use of supplementary sealers on gaskets. The gaskets are supplied as total sealing systems, and if properly installed on properly conditioned flanges, they do not require supplementary sealers.

The effects of temperature on gasket performance are very complex and not too well understood. In low-pressure applications, moderate temperatures appear to favor the initial seal; that is, it is improved so the joint becomes a little more impermeable to sealed fluids. This can be attributed to the softening effect produced in the gasket by initial heating. The gasket, being softened under loading conditions, will more than likely flow into the surface imperfections on the flange, thereby completing the conformation between gasket and flange. This is called "settling in."

Prolonged exposure to higher than ambient temperature will cause many gasket materials to harden. Fortunately, the hardening does not appear to seriously affect either the initial seal or the slight improvement in sealing caused by the initial heating.

While moderate temperatures promote sealing, abnormally high temperatures will result in a complete breakdown of the gasket. These are the temperatures which normally cause burning or charring in nonmetallic materials. Hence temperature can have both beneficial and detrimental effects on the initial seal. Therefore, the designer must avoid conditions which lead to the latter or must select gaskets which maintain their initial seals when exposed to such conditions.

Gasket materials, in general, have somewhat higher coefficients of thermal expansion than most of the metals from which flanges and bolts are made. In certain situations involving wide and rapid temperature fluctuations this factor of relative expansion and contraction due to such temperature changes may require special design considerations. The gasket must be able to seal when exposed to changing temperatures.

Even in joints where the flange pressures are high enough to produce initial seals, the internal fluid will penetrate the gasket to a slight degree. Such penetration, or edge effect, is perfectly normal and has little or no effect on the gasket's sealing ability. If anything, it aids sealing. Moderate swell may be highly beneficial even in assemblies where flange pressures are lower than those required for sealing. The swell will compensate for the lack of gasket loads and produce acceptable seals in joints which otherwise would exhibit leakage. On the other hand, excessive swell can be detrimental, particularly if the gasket becomes too soft and tends to disintegrate in the sealed liquid.

Shrinkage can also give rise to harmful effects. It is indicative of a loss of binder in the gasket. This means that, more than likely, the sealed fluid will become contaminated. Moreover, extraction or a loss of binder could very well lead to serious leakage.

Ultraviolet light, vermin, ozone, and fungi are all a part of a gasket's external environment. Their effect on the gasket may not be too well understood, but in some applications they can be objectionable. Ozone causes some rubber compounds to crack or split. Various fungicides, applied to the gasket in the form of surface treatments, will provide suitable resistance to mold growth.

When one considers the broad aspects of gasket performance or serviceability, obviously many factors or variables are involved. Those that fit into a seemingly logical pattern as far as sealing is concerned have been covered.

Other factors which tie into the many other aspects of gasket performance are electrical resistivity, thermal conductivity, gasket treatments, gasket tolerances, stability considerations, and resistance to abrasion.

Initial Seal Creation

The creation of the initial seal is the focal point for designing an effective gasket. The gasket must resist leakage through the material and have sufficient conformability with the mating flanges to result in sealing over the surface of the gasket. The gasket must have the macroconformability to conform to the flange distortion as well as the microconformability to flow into the surface finishes of the mating flanges. The clamping pressure on the gasket must be high enough for these to occur.

The clamping pressure on the gasket is a function of the number of fasteners, the torque imposed on them, the friction in the bolting system, and the compressed area of the gasket. The equations associated with these follow.

$$T = KDF \tag{2.1}$$

or

$$F = \frac{T}{KD} \text{ per fastener}$$

where T = torque, lb-in
D = nominal diameter of fastener, in
F = clamping force, lb
K = friction factor

$$F \text{ (total)} = \frac{NT}{KD} \tag{2.2}$$

where N = number of fasteners

$$P = \frac{F(\text{total})}{A} \tag{2.3}$$

where P = clamping pressure or stress on gasket, psi
A = compressed area of gasket, in^2

The frictional coefficient K can vary widely. It generally ranges from 0.16 for oiled fasteners to 0.20 for dry fasteners.

Coefficient values beyond these values can occur, especially with improved thread lubricants and/or galled threads, respectively. However, the 0.16 to 0.20 range does provide a good starting point for design purposes.

After determining the clamping pressure on the given gasket, one needs to check the sealability of the various gasket materials at this clamping pressure. The sealability of the material must be adequate to seal at the pressure of the media of the application. Gasket suppliers should be contacted in this regard if data are not available to the gasket design engineer. As can be expected, gasket materials tend to improve in sealing as the clamping pressure increases and as the internal pressure decreases. In addition, some materials tend to improve in sealing once the temperature of the application is reached as the gasket becomes seated.

The following are some of the properties of gasket materials that are important in the initial seal creation.

Conformability

In order to conform to the macrocompressibility requirements, which vary widely depending on the application, it is necessary to vary the load-bearing or stress-compression properties of the gasket in accordance with these conditions. Figure 2.16 illustrates stress versus compression curves for various gasket

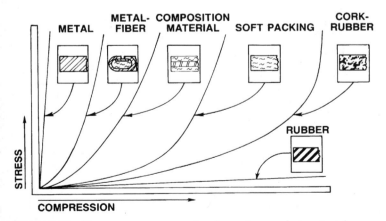

Figure 2.16 Stress versus compression for various gasket materials.

materials. In addition, adequate compression of the gasket must be achieved in order to be impervious to the media being sealed and conform to the required microcompressibility of the application. Depending upon the clamping pressure, different gasket materials may be selected. Selection procedure is discussed later.

Upon initial assembly, the bolts need to be capable of handling the maximum load imposed on them without yielding. In addition, the gasket must be capable of sealing at the clamping pressure imposed on it and also resist blowout at this load level.

Gaskets fabricated from compressible materials should be as thin as possible. The thickness should be no greater than that which is necessary for the gasket to conform to the unevenness of the mating flanges. The unevenness is associated with flange flatness and flange warpage during use. It is important to use the gasket's load curve when determining its ability to conform to the initial flange. The unload curve must be used when determining a gasket's ability to conform to the motion during operation. Figure 2.17 depicts typical load compression and unload curves for nonmetallic gasket materials.

The unload curve determines the recovery characteristic of the gasket, which is required for conformance. Load compression curves can be obtained from gasket suppliers.

Some advantages of thin gaskets over thick gaskets are:

1. Fewer voids through which sealing media can enter, therefore less permeability and better sealing within the material
2. Less distortion of mating flanges during assembly
3. Higher resistance to blowout
4. Reduced creep relaxation and subsequent torque loss
5. Reduced thickness tolerances
6. Better heat transfer
7. Lower cost

A common statement in the gasket industry is: "Make the gasket as thin as possible and as thick as necessary." Extremely thin gaskets, however, cannot sufficiently fill out excessive

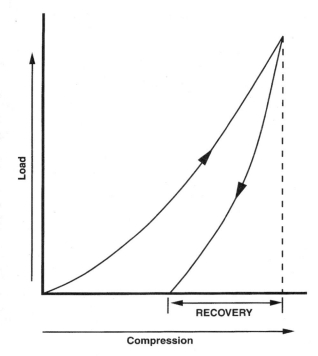

Figure 2.17 Load and unload curve for a typical non-metallic gasket material.

roughness depths or waviness to the extent that the flange pressure can prevent leakage of the pressurized agent. Therefore, when selecting gaskets, it is essential to check if the particular assembly might require a certain minimum gasket thickness.

Gasket thickness and compressibility must be matched to the rigidity, roughness, and unevenness of the mating flanges. A complete seal can be achieved only if the stress level imposed on the gasket at clamp-up is adequate for the specific material.

Stress-Distribution Testing

Most gasketed joints do not result in uniform stress distribution due to flange bending during assembly. A gasket designer needs to know the true stress distribution in order to determine whether or not a particular joint will be properly sealed with the selected gasket. A number of methods have been used to determine the stress distribution in a gasketed joint. These methods utilize the following:

1. Regular carbon paper
2. Lead pellets
3. Carbonless paper
4. Stress-sensitive film
5. Electrical sensors

Regular carbon paper

Regular carbon paper, as once extensively used by secretaries in typing, is a two-piece system. One sheet is the carbon carrier and the other is a clean sheet of paper required for carbon transfer. Stress impressions determined with this technique provide "on-off" or "yes-no" visual effect. That is, sufficient stress either was or was not available to transfer the carbon from the carbon paper to the clean sheet. This provided a very narrow range of stress information. In some instances multiple layers of carbon paper were used. In such cases, compressibility of the paper stack had to be taken into account, particularly with gaskets that inherently possessed little compressibility. Because of these limitations, regular carbon paper is rarely used today for stress-distribution determination.

Lead pellets

Lead shot or pellets is one of the earliest test methods associated with the clamping pressure on a gasket. It is still used today for determination of a gasket's compressed thickness at predetermined locations within a sealing joint. The test is also referred to as the "solder plug test." Since lead has virtually no recovery, its compressed thickness is the same as the gasket's compressed thickness under the clamp load of the application.

Lead pellets are placed into prepunched holes in the sealing gasket and the flange system is then assembled. See Fig. 2.18.

When the mating flanges are assembled, the pellets are compressed to the same thickness as the sealing gasket. Upon disassembly, the pellets are measured. See Fig. 2.19.

The Gasket and Its Environment   71

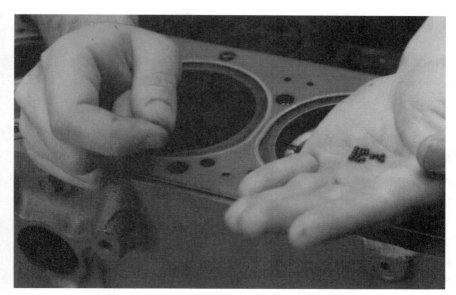

Figure 2.18 Lead pellet placement.

Figure 2.19 Lead pellet measurement.

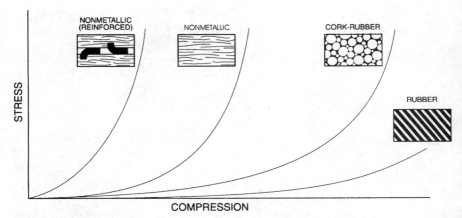
Figure 2.20 Stress versus compression for various gasket materials.

Using stress versus compression curves for the gasket material, the applied stress on the gasket can be determined at the various pellet locations. See Fig. 2.20.

Lead pellets, however, absorb load when compressed. The load absorbed reduces the load on the gasket. To minimize this effect, the smallest pellet diameter is chosen for the given gasket thickness. In many instances several pellets may be required to evaluate a sealing joint and considerable available load may be taken up by the lead pellets. See Fig. 2.21. This could affect the true sealing stress on the gasket in the actual application.

Various sections of ASTM F145, which is associated with lead pellet testing, follow:

ASTM F145-72 (reapproved 1987)

Standard practice for evaluating flat-faced gasketed joint assemblies. Scope: This practice permits measurement of gasket compression resulting from bolt loading on a flat-face joint assembly at ambient conditions.

Summary of Practice: The gasket compression and flange distortion are obtained from compressed-thickness measurements on cylindrically shaped soft-solder plugs (50-50 lead-tin by weight) inserted into holes, drilled or punched through the gasket in the thickness direction. Initial compression is accom-

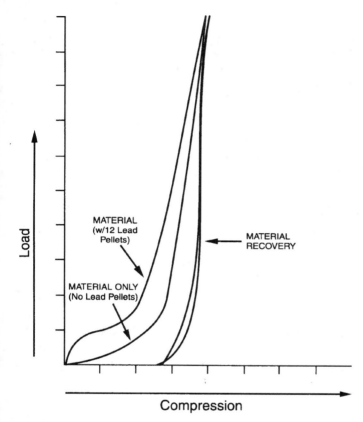

Figure 2.21 Load versus compression for gasket with and without lead pellets.

plished in the flanged-joint assembly when the bolts are loaded at ambient temperature. Solder, being inelastic, will remain at the compressed thickness of the gasket after the joint is subsequently disassembled.

Significance and Use: Gasket compressions produced by bolt loads in a flange joint are important in the application engineering of a joint assembly. They are related to the ability of a gasket to seal and to maintain tightness on assembly bolts, and to a variety of other gasket properties that determine the service behavior of a joint assembly. Thus, being able to determine the degree of compression in a gasket under the bolt loading will permit one to make qualitative predictions of the behavior of a joint assembly when it comes in contact with the application or service environment. With the plug test, bending of a flange facing between bolt centers can be measured; however, in a few highly distortable flanges the maximum bending between bolt centers may not be detected.

The variation in gasket compressions at selected points in a flat-face joint assembly reveals the degree of flange distortion or the ability of the flange to distribute satisfactorily the compressive forces from bolt loads throughout the gasket.

Apparatus:
- Test assembly, a flat-face flange design
- Torque-indicating device, for bolt loading
- Dial gauge indicator, graduated in 0.0001 in, to measure thickness of the solder plugs and the uncompressed gasket
- Leather punch, for punching holes in the gasket and fabricating the solder plugs
- Tweezers, to conveniently handle the solder plugs
- Solder plugs—The solder must be made into a flat strip. This can be done by compressing wire in a vise, a pair of flanges, or pliers, or passing it between two calender rolls. The solder plugs are punched from the strip by means of the leather punch. Recommended plug diameter is $\frac{1}{32}$ in, and the height need only be such that the plug is compressed by the flanges when the gasket is also compressed. The initial thickness of the plug and gasket before compression need not be equal.

Test Specimens: Three gasket specimens shall be tested. The size and shape of the specimens must be such as to fit the particular flange design.

Conditioning: When the test is performed on an assembly line or in a service environment, sufficient time should elapse for the flanges, bolts, and gasket to reach equilibrium with the ambient temperature and humidity conditions before assembly. (Heavy castings or forgings may require 8 to 24 h or more, contrasted to a brief period for light stampings.)

When the test is performed under controlled conditions in the laboratory, the gasket specimen is conditioned in accordance with Specification E171, or in the humidity and temperature conditions used prior to obtaining the load-compression curve. Flange fasteners and washers are held at the test conditions for at least 4 h prior to assembly.

Procedure: Use the same flanges, fasteners, and washers as those specified for applications. Clean them with reagent-grade trichloroethylene or other suitable solvent. Use cleansing tissue to remove dirt, oil, or grease. After cleaning, give the mating screw threads a light coating on SAE 20 engine oil to minimize friction.

Measure the initial or uncompressed thickness of the gasket. Make holes, slightly larger than the solder plugs, in the gasket at the points where compressions are to be evaluated. Insert the plugs upright in the holes with the gasket resting on the lower flange facing. Then assemble the test assembly in the customary manner. Immediately disassemble the test assembly. Make compressed-thickness measurements on the solder plugs. These measurements are equal to the compressed thickness exhibited by the gasket when it was loaded in the test assembly.

Calculation and Interpretation of Results: Calculate the compression as a percentage of the original gasket thickness as follows:

$$C = \frac{t_0 - t_c}{t_0} \times 100$$

where C = percentage compression in the gasket
t_0 = initial uncompressed thickness of the gasket
t_c = compressed-gasket thickness as measured on a solder plug

The plug test will indicate the distribution of gasket compressions which reflects the variation of initial flange pressure in joint assemblies exhibiting flange distortion.

Report: The report shall include the following:
- Conditions of test, temperature, relative humidity, and time of conditioning
- Gasket identification

- Conditioning of gasket before test
- Uncompressed-gasket thickness
- Flange; material, thickness, width, and bolt spacing
- Bolt size, material, and thread condition
- Method of bolt loading
- Bolt torque
- Tightening sequence on bolts
- Top view of gasket showing solder-plug locations with location dimensions
- Compressed-gasket thickness at each plug location
- Percentage compression in the gasket at each plug location
- Data plot

Carbonless paper

Carbonless (no carbon required) paper, also known as no carbon required (NCR) paper, is a uniquely coated paper made to produce an image from mechanical pressure. An image is created when two separate colorless chemical compounds previously coated on the sheets are brought together. These compounds or chemical dyes are separately sealed within millions of plastic-like microscopic capsules. Figure 2.22 depicts the NCR paper makeup.

After being applied onto the carrier paper, these capsules are covered with a chemical coreactant which is dry to the touch. When pressure is applied to the paper, the capsules are broken,

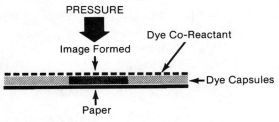

Figure 2.22 NCR paper makeup.

allowing the chemicals (dye and coreactant) to combine to produce an image. The number of broken capsules is proportional to the applied stress. A stress impression of a gasket using carbonless paper is shown in Fig. 2.23.

This paper is an improvement over regular carbon paper, since it is less intrusive, and the impression density or color intensity is proportional to the stress applied. Therefore, incremental impressions at various stress levels, performed on a tensile–compression load frame, can be used to calibrate the paper color impression density versus stress level (Fig. 2.24). Since the paper is available in only one pressure range, it provides limited quantitative data for applications where a broad stress range is present.

Stress-sensitive film

Stress-sensitive film, manufactured by Fuji Photo Film Co., Ltd., is available in one- or two-sheet systems and in three or four pressure ranges. It functions in a manner similar to carbonless paper. This film is an improvement over carbonless paper in that it permits "fine tuning" of the image to reflect stress distribution. There is a densitometer on the market which can be used in conjunction with Fuji film. It measures

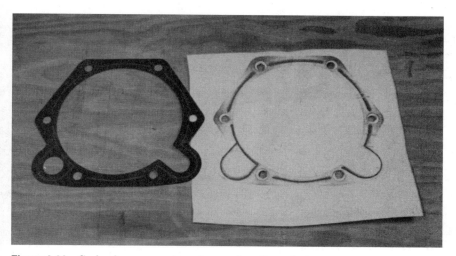

Figure 2.23 Carbonless paper stress impression of a gasket.

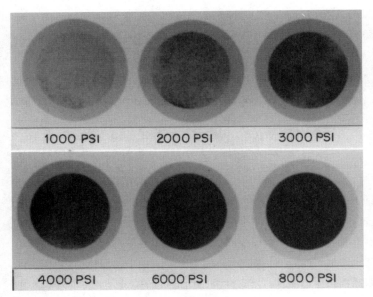

Figure 2.24 Carbonless paper calibration.

the stress levels of the film. Using the commercially available Fuji densitometer, impression color density can be directly converted to stress readings (Fig. 2.25). The film image is affected by time, temperature, and humidity, and these effects must be taken into account when analyzing color intensities.

As with carbonless paper, calibration of Fuji prescale film can be accomplished by performing incremental compressive-stress tests. This is done by applying a known compressive stress to the four grades of film (Fig. 2.26).

To perform stress-distribution evaluations in a gasketed joint, regular carbon paper, carbonless paper, or Fuji prescale film is precut to the shape of the mating flanges and holes are prepunched for the fasteners. The paper or film is then placed between the flanges and the fasteners tightened to a specified torque. When the paper is removed from the actual gasketed joint, it shows the stress-distribution pattern on the gasket (Fig. 2.27).

By using the comparison sheet that has been preloaded to various stress levels the impression from the actual gasketed joint can be visually compared to obtain the applied stress at

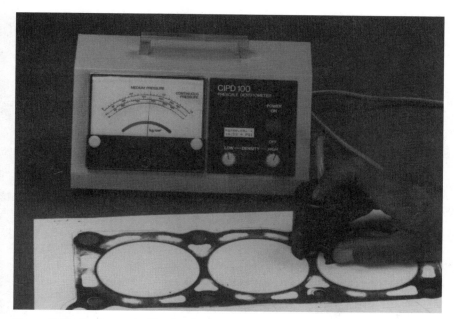

Figure 2.25 Fuji prescale film densitometer stress impression reading.

various gasket locations. The gasket engineer uses the stress-distribution data to determine if the gasket will seal. If this stress distribution is inadequate for good sealing, a revised gasket material or design is needed.

A deficiency in each of the above techniques is that they all indicate a one-time maximum stress during the course of testing. Any reduction in stress during the torquing sequence or from external forces is not reflected in the impressions. The gasket designer needs to be aware of this, especially in low-clamp-load and nonrigid flange joints.

Electrical sensor construction and technology

There is new technology which can be used to measure stress distribution during both loading and unloading. It is described in the ASME/JSME paper, "Measuring Real Time Static and Dynamic Gasket Stress Using a New Technique," June 1991, Book H00644. The American Society of Mechanical Engineers

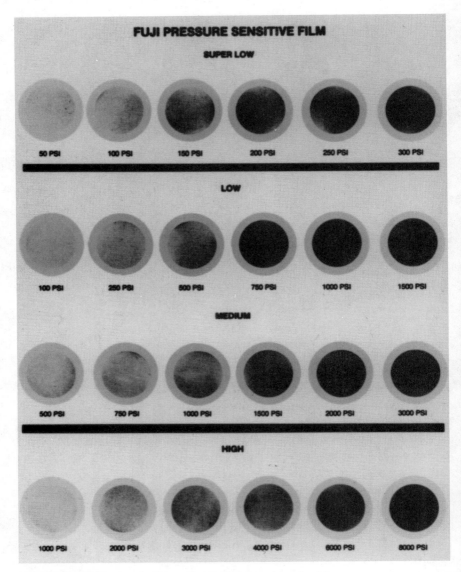

Figure 2.26 Fuji prescale film calibration made using a load frame.

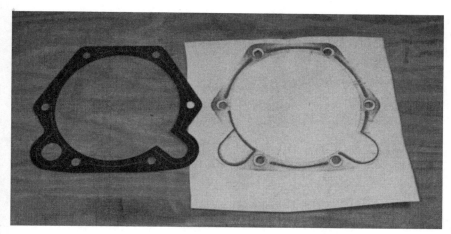

Figure 2.27 Gasket and stress distribution on it.

is located at 345 East 47th St., New York, NY 10017. The technology is briefly discussed here.

The sensor is approximately 0.004 in thick. It consists of two flexible, thin polyester sheets which have electrical strip patterns (electrodes) deposited on them. The inside surface of one sheet has a row pattern while the inside surface of the other sheet has a column pattern. The spacing between the strip patterns (rows and columns) can vary down to a minimum of approximately 0.040 in. When the two sheets are placed one on top of the other, they form a grid pattern providing a sensing cell at each intersection.

Before assembly, a thin semiconductive coating (ink) is applied as an intermediate layer between the electrical contacts (rows and columns) (see Fig. 2.28). This ink provides an electrical resistance and conductance between intersecting contacts.

Of major importance is the ink's electrical resistance, which changes with applied external force. Therefore, when installed with a gasket and clamped within the sealing joint, the sensor provides an array of pressure-sensitive cells.

By measuring the minute changes in current flow at each intersecting point, the applied force distribution pattern can be measured and displayed on a computer screen. This pattern shows the location and magnitude of the forces exerted on the

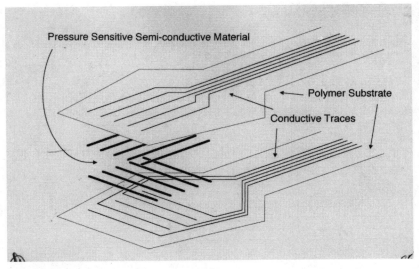

Figure 2.28 Electrical sensor construction.

surface of the sensor at each cell. Software has been developed to permit two- and three-dimensional color displays to show the relative force at each cell. Changes in force can be observed, measured, and recorded throughout the test. This provides a powerful gasket design engineering tool.

The semiconductive ink contains conductive particles suspended in a polymer-based binder. As the particles are brought closer together with applied force, the electrical resistance through the ink is reduced.

The use of any one of the compressive-stress-measurement methods (carbon paper, carbonless paper, Fuji prescale film, and electrical sensor) adds thickness as well as their own deflection characteristics. To minimize the effects of the sensor on the sealing system, it is desirable that the sensor deflect as little as possible within its operating load range. As shown in Fig. 2.29, the conditioned (preloaded) electrical sensor deflects the least. The other techniques, of course, don't permit preloadings.

Computer hardware and software. Hardware and software, along with a specialized electrical contact handle, is used in conjunction with a personal computer to display the output from the sensor. In the display is a color bar chart which depicts the

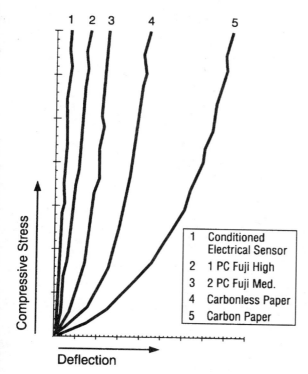

Figure 2.29 Compressive stress versus deflection for the various measurement tools.

relative amount of stress at each sensing cell. In addition, the current technology includes a sum force scale that indicates relative differences in load levels at the various points. Also designed into the software is a windowing feature that allows localized area stress evaluation. Various types of universal and customized sensors are utilized with this measuring system. Sensor configuration design can be made to fit most application requirements. Figure 2.30 depicts a number of sensors that have been used.

A component diagram of the sensing system is shown in Fig. 2.31. The sensing system is controlled by a personal computer. The menu-driven software is controlled from the computer keyboard, and the output is displayed on the color monitor and/or printer. See Fig. 2.32.

84   Chapter Two

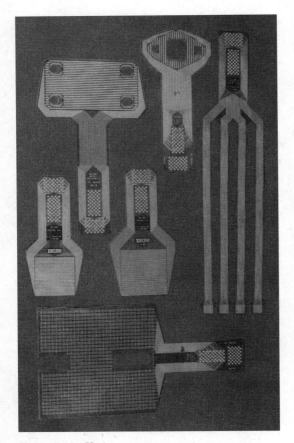

Figure 2.30 Various sensors.

In designing flanged gasketed joints, the more that is known and understood about the behavior of stress distribution under dynamic conditions, the better a job can be done in gasket design. Many gasketed joints unload from hydrostatic end forces. This new technique provides information necessary to determine stress distribution in the sealing joint from static and/or dynamic unloading conditions.

Figure 2.33 shows the stress-distribution range that each of the techniques is capable of determining. Note the large capability of the electrical sensor technology.

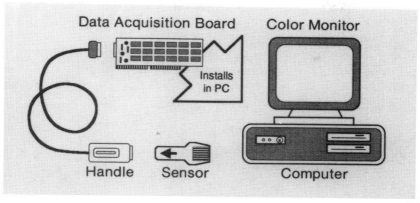

Figure 2.31 Component diagram of the sensing system.

Figure 2.32 Output display of the electrical sensing system.

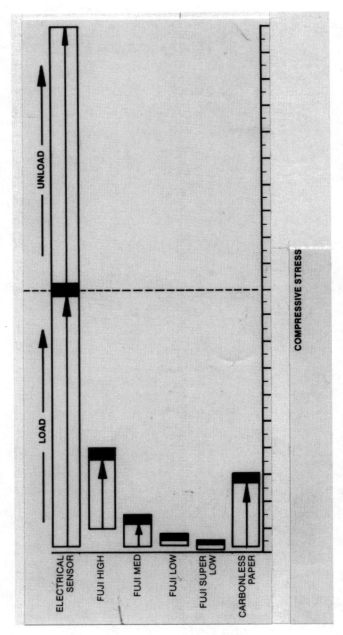

Figure 2.33 Output comparison between the various measurement tools.

References

Brink, Czernik, and Horve: *Handbook of Fluid Sealing,* McGraw-Hill, New York, 1993.

Czernik, D. E., and Frank L. Miszczak: "Measuring Real Time Static and Dynamic Gasket Stresses Using a New Technique," ASME/JSME paper, 910205, June 1991.

Czernik, D. E., J. C. Moerk, Jr., and F. A. Robbins: "The Relationship of a Gasket's Physical Properties to the Sealing Phenomena," SAE paper 650431, May 1965.

Lindeman, C. M., and R. D. Andrew: "Fibers for Consideration in Non-Asbestos, High Performance Gasket Materials," SAE paper 810364, presented at the SAE International Congress and Exposition, February 1981.

SAE AE-13: "Gasket and Joint Design Manual for Engines and Transmission Systems," 1988.

Tracy, D. H., and B. M. Arnio: "Performance and Reliability of Non-Asbestos Gasketing Materials," SAE paper 830218, presented at the SAE International Congress and Exposition, February 1983.

Chapter 3

The Gasket and the Joint

Maintenance of the Seal

After the initial sealing stress is applied to a gasket, it is necessary to maintain a sufficient stress for the designed life of the unit. This is most important for maintenance of the seal. All materials exhibit, in varying degrees, a decrease in applied stress as a function of time, commonly referred to as stress relaxation. The reduction of stress on a gasket is actually a combination of two major factors, stress relaxation and creep (compression drift). By definition:

Stress relaxation is a change in stress s on a gasket under constant strain e (ds/dt; e = constant).

Creep (compression drift) is a change in strain of a gasket under constant stress s (de/dt; s = constant).

In a gasketed joint, stress is applied by tension in a bolt or stud and transmitted as a compressive force to the gasket. After loading, stress relaxation and creep occur in the gasket, causing corresponding lower strain and tension in the bolt. This process continues indefinitely as a function of time. The change in tension of a bolt is related to the often-quoted "torque loss" associated with a gasket application. Since the change in stress is due to two primary factors, a more accurate description of the phenomenon would be creep relaxation, hereafter called relaxation for simplicity purposes. The basic result is a loosening of the gasketed joint and a tendency for leakage.

Some of the factors which are influential in this regard are temperature and mechanical related. Thermal expansion in joints using steel fasteners and aluminum flanges can be a major contributor to relaxation. Aluminum flanges expand more upon heating than do the fasteners. This results in increased clamping pressure and squeeze on the gasket. Upon cooling, the aluminum flange contracts more than the steel fasteners and a reduction of clamping pressure occurs. The gasket must have adequate recovery to compensate for the difference in contraction. Relaxation of the gasket due to the additional squeeze and subsequent reduction of stress results in relaxation.

In regard to the mechanical aspects, the dynamic impact on the gasket due to the hydrostatic end force results in a compressed thickness reduction of the gasket and subsequent clamping pressure loss. Again, the gasket's recovery characteristics must be capable of handling this change in compressed thickness. Therefore, owing to the relaxation associated with thermal and mechanical effects, the maintaining of the seal is affected.

Relaxation

A loss in thickness of a gasket which results in relaxation directly results in loss of stretch in the clamping bolts. Bolt elongation, or stretch, is linearly proportional to bolt length. The longer the bolt, the higher the elongation. The higher the elongation, the lower the percentage loss for a given relaxation. Therefore, the bolts should be made as long as possible for best torque retention.

Relaxation is defined in this handbook as the amount of retained stress divided by the original stress on the gasket. Relaxation in a gasket material may be measured by applying a load on a specimen by means of a strain-gauged bolt-nut-platen arrangement as standardized by ASTM F38-62T.

The ASTM F38 standard test device is depicted in Fig. 3.1. Selection of materials with good relaxation properties will result in the highest retained torque for the application. This

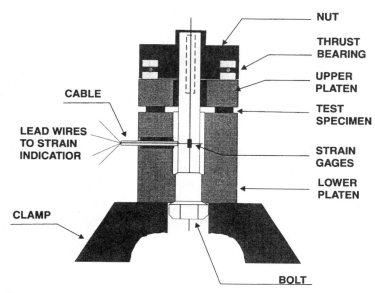

Figure 3.1 Schematic diagram of the ASTM relaxometer.

results in the highest remaining stress on the gasket which is desirable for long-term sealing performance.

The amount of relaxation increases as gasket material thickness is increased. This is another reason why the thinnest gasket that will seal should be selected. Figure 3.2 depicts the relaxation characteristics as a function of thickness and stress level for a particular gasket material.

Relaxation is decreased as clamping stress is increased. This is a result of closing the voids in the material as stress is increased (Fig. 3.3).

Relaxation is normally reduced as the density of the gasket is increased. This is true if the percentage of binder is not increased (see Fig. 3.4).

Relaxation has been determined to be of loglog function versus time. Figure 3.5 shows the relaxation versus time on a loglog plot for three different gasket materials. When the same data are plotted on semilog paper, a slight nonlinear section is noted (see Fig. 3.6). In a gasketed application this will result in

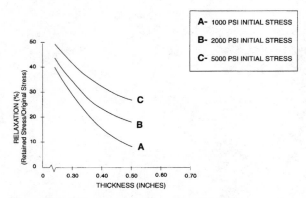

Figure 3.2 Relaxation versus gasket thickness.

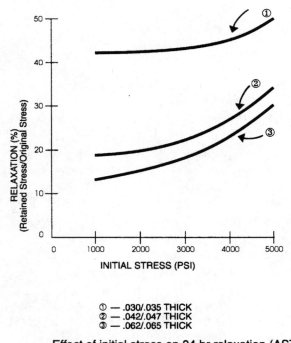

Effect of initial stress on 24 hr relaxation (ASTM F38)

Figure 3.3 Relaxation versus initial clamping stress.

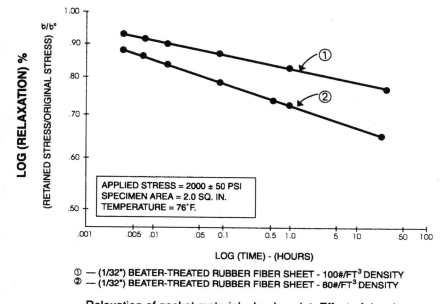

Figure 3.4 Effect of density on relaxation.

a large decrease in clamp-up load immediately after assembly and then a leveling off of the load.

Effects of Unit Operation

The stress-compression and relaxation characteristics of gasketing materials do not remain constant with exposure to operation. In general, the materials exhibit higher spring rate and lower relaxation after operation, tending to approach a maximum value of spring rate after a certain number of hours of operating time. Various materials have been tested to determine their stress-compression properties after subjection to operation. Figure 3.7 shows some typical results of increased spring rates. In order to simulate the effect of operation, samples were also subjected to repeated cyclic loading by means of an Instron universal testing machine; these effects are also plotted in Fig. 3.7.

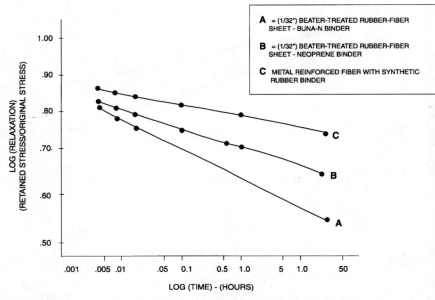

Figure 3.5 Relaxation of gasket materials, loglog plot.

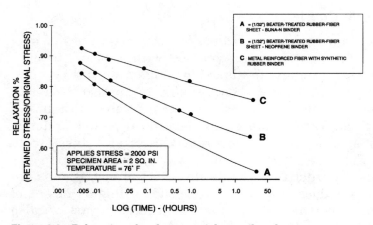

Figure 3.6 Relaxation of gasket materials, semilog plot.

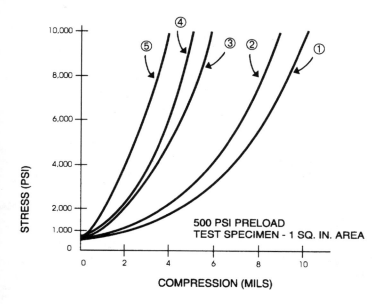

Figure 3.7 Stress versus compression, effect of load applications.

Effects of Temperature

The stress-compression properties are, owing to their viscoelastic properties, also a function of temperature. Figure 3.8 shows a typical change in stress-compression properties of samples tested at room temperature and at 300°F.

The change in the gasket's spring rate properties has been associated with the torque loss existent in the application, as previously mentioned. The association of the relaxation characteristics with torque loss is discussed later.

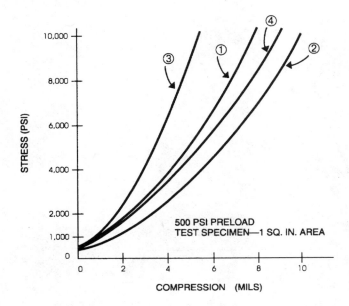

Stress versus compression: Effect of temperature and load application

Figure 3.8 Stress versus compression, effect of temperature.

Gasket Shape Factor

The gasket's plan view and thickness have an effect on its relaxation characteristics. This is particularly true in the case of the more highly compressible materials.

Some of the relaxation of a material may be attributed to the releasing of forces through lateral expansion or bulging. Therefore, the greater the area available for lateral expansion, the greater the relaxation. The shape factor of a gasket is the ratio of the area of one load face to the area free to bulge. For circular or annular samples, this may be expressed as

$$\text{Shape factor (S.F.)} = \frac{\text{area of one load face}}{\text{area free to bulge}}$$

$$= \frac{(\pi/4)[(OD)^2-(ID)^2]}{\pi(OD + ID)h}$$

$$\text{Shape factor} = \frac{(OD + ID)(OD-ID)}{4h(OD + ID)}$$

$$= \frac{1}{4h}(OD-ID)$$

where h = height (thickness) of gasket
 OD = outer diameter
 ID = inner diameter

As the area free to bulge increases, the shape factor decreases, the relaxation increases, and the retained stress decreases. Figure 3.9 depicts the effect of shape factor on the gasket's ability to retain stress.

As can be noted in the shape factor equation, the shape factor decreases with increasing thickness. This is another reason why the gasket should be as thin as possible. As noted earlier, it must be thick enough, however, to permit adequate conformi-

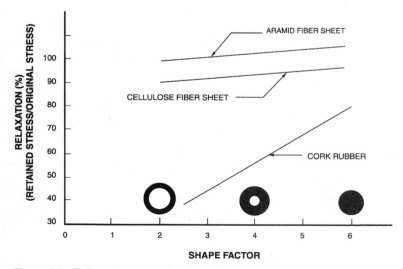

Figure 3.9 Relaxation versus the shape factor.

ty. The clamp area should be as large as possible consistent with sealing stress requirements. Often designers reduce gasket width, and subsequent clamp area, thereby increasing gasket clamping stress to obtain better sealing. Remember, however, this reduction will decrease the gasket's shape factor, which may result in higher relaxation. Therefore, a compromise must be made when a gasket's clamp area is reduced to retain adequate sealing stress.

Relaxation and Torque Loss

In most gasket applications, a quantity called "torque loss" is widely used as a criterion for predicting gasket performance. A gasket which has low torque loss is preferred to one which has high torque loss.

The reasoning behind this thinking is simple—torque on a bolt is related to tension of that bolt and the tension in the bolt is related to the compressive or sealing stress existent on the gasket which is clamped by the bolt. The loss, as noted earlier, is the relaxation. Therefore, if torque is related to tension which in turn is related to gasket stress, then torque must be related to gasket stress. A loss of torque therefore means a loss in gasket sealing stress.

This thinking, although simple in theory, is rather complex in actuality. The main reason for this complexity is friction. Owing to friction, torque and tension are usually not directly related. The amount of tension existent in a bolt for a given torque is dependent upon many factors—two of the main ones are (1) friction between the bearing surface of the bolt head and flange and (2) friction in the threads. Unfortunately, these frictions are not constant from one application to the next—or even constant in the same application after repeated use of the same bolt in the same hole.

There is one consolation, however, inherent in a bolted assembly that allows the conversion of torque to tension with realistic accuracy. This consolation is that about 85 to 90 percent of the torque applied to a bolt is "used up" in overcoming friction and the remaining 10 to 15 percent is actually used in elongating the bolt. This high friction percentage appears at a cursory glance to be a hindrance rather than a consolation. However, upon further investigation, one realizes that if the friction does change in any particular application, the tensile change associated with the change in friction is far less percentagewise. Theory on this follows later. This means, if the friction does change, that within limits one can associate torque to clamping stress and "torque loss" to loss in clamping stress.

One noteworthy point to interject at this time, however, is that "torque loss" is only one factor involved in measuring a gasket's performance. There have been actual situations in which a loss in torque resulted in a redistribution of loading, which aided sealing.

Torque and torque loss

Torque can be defined as the tendency of a force to produce rotation about an axis. The units for torque are the same as the units for energy, viz., foot-pounds or inch-pounds. A torque wrench should be used for gasket assembly. Figure 3.10 shows one type of torque wrench.

"Torque loss" can be defined as the difference in the initial torque and the pull-up torque. Pull-up torque is measured by (1) scribing the relative positions of a bolt and flange, (2) backing the bolt off approximately 90°, and (3) measuring the torque required to reposition the bolt to the original scribed lines. Some equipment manufacturers do not back off the bolt

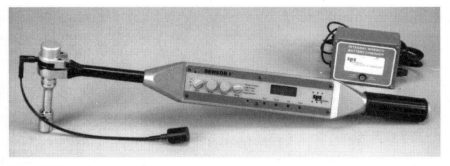

Figure 3.10 A torque wrench.

but rather apply torque in the tightening direction and measure the retained torque by noting the torque just at the instant the bolt begins to move. The difference in the initial torque and the noted torque is their measure of torque loss. The former method is preferred, since it produces more consistent results.

Tension and tension loss

Tension is the pull exerted on the bolt. The units of tension are the same as those of weight, viz., pounds. The measurement of tension in a bolt is generally determined by actually measuring the stretch of the bolt which is caused by the tension. The tension is directly proportional to this stretch. Various techniques have been used to measure bolt stretch. Some of these are:

1. Direct measurement with micrometer. Many times, of course, this is impossible in actual applications.
2. Angular rotation measurement. Since bolt threads operate on a helix angle, it is possible to determine stretch based on the angular rotation. This method is limited by any deformation that occurs in the mating threads or the bearing surfaces.
3. Gauge pin bolt. This is a bolt which has a gauging pin located along the longitudinal axis of the bolt and fixed at the bottom. Although the bolt undergoes stretch, the pin does not, and there is a relative displacement between the pin and the bolt. This displacement is a measure of the bolt's elongation.
4. Strain-gauged bolts. A strain gauge is a thin metallic resistive element which changes its resistance in proportion to its strain. These gauges are cemented to bolts and undergo the same strain that the bolt undergoes when it is loaded. By measuring the change in resistance and having knowledge of the relationship between resistance and strain, one can relate the change in resistance to bolt stretch.
5. Ultrasonic measurements. A new technology involves the ultrasonic measurement of the change in length of bolts and is expected to make stretch control far more common.

The method uses acoustic energy, and an instrument measures the time for the acoustic wave to travel from one end of the bolt and return. As the bolt is stretched, the time for the wave to complete its path is increased.

The relationship between torque loss and tension loss, although highly dependent upon friction, can be used in gasket applications for comparative purposes. Torque loss results should be used only as an average, however, and any torque loss readings significantly different from the mean should not be construed to mean significantly different tension losses. Using the methods indicated previously, it has been found that average torque losses and average tension losses are accurate in many applications to within ± 10 percent. In some applications, however, ± 20 percent variations have been encountered.

Tension versus torque (effect of friction)

$$T = KDP$$

where

$$K = u_B \frac{R_B}{D} + u_T \frac{R_T}{D} \sec B + \frac{R_T}{D} \tan C*$$

or

$$K = K_1 + K_2 + K_3$$

where K_1 = unit torque wasted by friction on bearing face of nut or bolt
K_2 = units torque wasted by friction on contact flanks of threads
K_3 = useful torque producing bolt tension

Symbols

T = torque, lb-in
K = torque coefficient
D = nominal bolt diameter, in
P = bolt tension, lb

---

*"Fasteners Book," *Machine Design*, Sept. 29, 1960.

u_B = coefficient of friction at bearing face of nut or bolt
u_T = coefficient of friction at thread contact surfaces
R_B = effective radius of action of frictional forces on bearing face, in
R_T = effective radius of action of frictional forces on thread surfaces, in
B = thread half angle
C = helix angle of thread
d = thread depth

Case 1

$$U_B = U_T = 0.15 \qquad \tan C = 0.04$$

$$\frac{R_B}{D} = 0.65 \qquad \frac{R_T}{D} = 0.45 \qquad \sec B = 1.15$$

$$K_1 = u_B \frac{R_B}{D}$$

$$K_1 = 0.15 \times 0.65 = \underline{0.098}$$

$$K_2 = U_T R_T \sec B$$

$$K_2 = 0.15 \times 0.45 \times 1.15 = \underline{0.078}$$

$$K_3 = \frac{R_T}{D} \tan C$$

$$K_3 = 0.45 \times 0.04 = 0.018$$

$$K_1 + K_2 + K_3 = 0.098 + 0.078 + 0.018 = 0.194$$

$$\text{Percent } \frac{K_1}{K} = \frac{0.098}{0.194} \times 100 = 50.5 \text{ percent}$$

$$\text{Percent } \frac{K_2}{K} = \frac{0.078}{0.194} \times 100 = 40.2 \text{ percent}$$

$$\text{Percent } \frac{K_3}{K} = \frac{0.018}{0.194} \times 100 = 9.3 \text{ percent}$$

This means 9.3 percent of the applied torque is used to produce tension in the bolt.

Case 2. All parameters are the same as Case 1 except $u_B = u_T = 0.1$. This represents a change in frictional coefficient of $0.5/0.15 \times 100 = 33$ percent. Now:

$$K_1 = 0.10 \times 0.65 = 0.065$$

$$K_2 = 0.10 \times 0.45 \times 1.15 = 0.0517$$

$$K_3 = 0.45 \times 0.04 = 0.018$$

$$K_1 + K_2 + K_3 = 0.065 + 0.0517 + 0.018 = 0.1347$$

$$\text{Percent } \frac{K_1}{K} = \frac{0.065}{0.1347} \times 100 = 48.3 \text{ percent}$$

$$\text{Percent } \frac{K_2}{K} = \frac{0.0517}{0.1347} \times 100 = 38.3 \text{ percent}$$

$$\text{Percent } \frac{K_3}{K} = \frac{0.018}{0.1347} \times 100 = 13.4 \text{ percent}$$

In this case 13.4 percent of the applied torque is used to produce tension in the bolt.

Therefore, even though the friction coefficient changed by 33 percent, the percent change in tension for a given torque was 4.1 percent. In reality, the frictional change is usually higher than 33 percent, and therefore the tension change would be greater also. This situation was presented, however, to indicate the consolation mentioned in the text.

Actual measurement of percent torque used to produce tension

An actual case in a gasket application investigated is the following: For a $\frac{9}{16} - 12$ bolt, a torque of 140 lb-ft resulted in 18,800 lb of tension. Now:

$$K_3 = R_T \tan C$$

where $R_T = \dfrac{D_0 + D_i}{4} = \dfrac{D_0 + (D_0 - 2d)}{4}$

$$R_T = \dfrac{0.5625 + (0.5625 - 0.1080)}{4}$$

$R_T = 0.254$ in

$D = \dfrac{9}{16} = 0.5625$ in

$C = 2.6°$
$\tan C = 0.0507$

Thus

$$K_3 = \dfrac{0.254 \times 0.0507}{0.5625} = 0.0229$$

and

$$T_3 = K_3 DP$$

where T_3 is the torque that goes into producing tension P in the bolt,

$$T_3 = (0.0229)(0.5625 \text{ in})(1.88 \times 10^4 \text{ lb})$$

$$T_3 = 242.0 \text{ in-lb} = 20.15 \text{ ft-lb}$$

Percent of torque going into tension $= \dfrac{T_3 \times 100}{T}$. Therefore,

$$\text{Total} = \dfrac{(20.15 \text{ ft-lb}) \times 100}{140 \text{ ft-lb}}$$

$$= 14.4 \text{ percent, which is the range of}$$
$$10 \text{ to } 15 \text{ percent noted previously}$$

Joint and Gasket Design Selection

There are two commonly used design procedures to determine the necessary bolt loads to seal various gaskets. These procedures are associated with gasketed joints which have rigid, usually cast-iron flanges, have high clamp loads, and generally contain high pressures. Many gasketed joints have stamped-metal covers and splash or very low fluid pressure. In these cases, the procedures do not apply and stress-distribution testing for selection of appropriate gaskets should be considered by the designer.

Joint selection

The first step in the selection of a gasket for sealing a specific application is to choose a material that is both chemically compatible with the medium being sealed and thermally stable at the operating temperature of the application. The remainder of the selection procedure is associated with the minimum initial seating stress of the gasket and the operational stress on the gasket during operation. In these regards, two methods are reviewed. They are:

1. The American Society of Mechanical Engineers (ASME) Code Method
2. The simplified procedure proposed by Whalen

Table 3.1 depicts some typical gasket designs used for ASME code applications.

ASME code procedure

The ASME Code for Pressure Vessels, Sec. VIII, Div. 1, App. 2, is the most commonly used design method for gasketed joints. It should be noted that the ASME is currently evaluating the Pressure Vessel Research Council (PVRC) method for gasket design selection. It is likely that a nonmandatory appendix to the code will first appear. The PVRC method is discussed later.

An integral part of the ASME code centers on two gasket factors:

1. An m factor, often called the gasket maintenance factor, which is associated with the hydrostatic end force and the operation of the joint.

TABLE 3.1 Typical Gasket Designs and Descriptions

TYPICAL GASKET DESIGNS AND DESCRIPTIONS

Type	Cross-Section	Comments
FLAT		Basic form. Available in wide variety of materials. Easily fabricated into different shapes.
REINFORCED		Fabric or metal reinforced. Improves torque retention and blow-out resistance of flat types. Reinforced type can be corrugated.
FLAT WITH RUBBER BEADS		Rubber beads located on flat or reinforced material affords high unit sealing pressure and high degree of conformability.
FLAT WITH METAL GROMMET		Metal grommet affords protection to base material from medium and provides high unit sealing stress. Soft metal wires can be put under grommet for higher unit sealing stress.
PLAIN METAL JACKET		Basic sandwich type. Filler is compressible. Metal affords protection to filler on one edge and across surfaces.
CORRUGATED OR EMBOSSED		Corrugations provide for increased sealing pressure and higher conformability. Primarily circular in shape. Corrugations can be filled with soft filler.
PROFILE		Multiple sealing surfaces. Seating stress decreases with increase in pitch. Wide varieties of designs are available.
SPIRAL-WOUND		Interleaving pattern of metal and filler. Ratio of metal to filler can be varied to meet demands of different applications.

2. A y factor, which is the minimum seating stress associated with particular gasket material. The y factor is concerned only with the initial assembly of the joint.

The m factor is essentially a multiplier on pressure to increase the gasket clamping load to such an amount that the hydrostatic end force does not unseat the gasket to the point of leakage. The factors were originally determined in 1937, and even though there have been objections to their specific values, they have remained essentially unchanged to date. The values are only suggestions and are not mandatory.

This method uses two basic equations for calculating required bolt load, and the larger of the two calculations is used for design. The first equation is associated with W_{m2} and is the required bolt load to initially seat the gasket:

$$W_{m_2} = (3.14)bGy \tag{3.1}$$

The second equation states that the required bolt operating load must be sufficient to contain the hydrostatic end force and simultaneously maintain adequate compression on the gasket to ensure sealing:

$$W_{m_1} = \frac{3.14}{4} G^2 P + 2b(3.14)G_{mP} \tag{3.2}$$

where W_{m_1} = required bolt load for maximum operating or working conditions, lb
W_{m_2} = required initial bolt load at atmospheric temperature conditions without internal pressure, lb
G = diameter at location of gasket load reaction, generally defined as follows: When b_0 is less than or equal to ¼ in, G = mean diameter of gasket contact face, in; when b_0 is greater than ¼ in, G = outside diameter of gasket contact face less $2b$, in
P = maximum allowable working pressure, psi
b = effective gasket or joint contact surface seating width, in

$2b$ = effective gasket or joint contact surface pressure width, in

b_0 = basic gasket seating width per Table 3.2 (the table defines b_0 in terms of flange finish and type of gasket, usually from one-half to one-fourth gasket contact width)

m = gasket factor per Table 3.2 (the table shows m for different types and thicknesses of gaskets ranging from 0.5 to 6.5)

y = gasket or joint contact surface unit seating load, psi (per Table 3.2, which shows values from 0 to 26,000 psi)

Tables 3.2 and 3.3 are reprints of Tables 25-1 and 25-2 of the 1980 ASME code.

To determine bolt diameter based on required load and a specific torque for the grade of bolt, the following is used:

$$W_b = \frac{T}{0.17D} \quad \text{(for lubricated bolts)} \tag{3.3}$$

or

$$W_b = \frac{T}{0.2D} \quad \text{(for unlubricated bolts)} \tag{3.4}$$

where W_b = load per bolt, lb
D = bolt diameter, in
T = torque for grade of bolt selected, lb-in

Note that W_b is the load per bolt and must be multiplied by the number of bolts to obtain total bolt load.

To determine the bolt diameter based on the required load and the allowable bolt stress for a given grade of bolt, use

$$W_b = S_b A_b \tag{3.5}$$

where W_b = load per bolt, lb
S_b = allowable bolt stress for grade of bolt selected, psi
A_b = minimum cross-sectional area of bolt, in^2

TABLE 3.2 Gasket Materials and Contact Facings

Gasket Materials and Contact Facings†
Gasket Factors m for Operating Conditions and Minimum Design Seating Stress y

Gasket material	Gasket factor m	Minimum design seating stress y, psi	Sketches	Facing sketch and column to be used from Table 26-4
Elastomers with cotton fabric insertion	1.25	400		(1a), (1b), (1c), (1d), (4), (5); column II
Elastomers with asbestos fabric insertion (with or without wire reinforcement):				
3-ply	2.25	2 200		(1a), (1b), (1c), (1d), (4), (5); column II
2-ply	2.50	2 900		
1-ply	2.75	3 700		
Vegetable fiber	1.75	1 100		(1a), (1b), (1c), (1d), (4), (5); column II
Spiral wound metal, asbestos-filled:				
Carbon	2.50	10 000		(1a), (1b); column II
Stainless or Monel	3.00	10 000		

Gasket Materials and Contact Facings†
Gasket Factors m for Operating Conditions and Minimum Design Seating Stress y

Gasket material	Gasket factor m	Minimum design seating stress y, psi	Sketches	Facing sketch and column to be used from Table 26-4
Corrugated metal, asbestos inserted or corrugated metal, jacketed asbestos-filled:				
Soft aluminum	2.50	2 900		(1a), (1b); column II
Soft copper or brass	2.75	3 700		
Iron or soft steel	3.00	4 500		
Monel or 4-6% chrome	3.25	5 500		
Stainless steels	3.50	6 500		
Corrugated Metal:				
Soft aluminum	2.75	3 700		(1a), (1b), (1c), (1d); column II
Soft copper or brass	3.00	4 500		
Iron or soft steel	3.25	5 500		
Monel or 4-6% chrome	3.50	6 500		
Stainless steels	3.75	7 600		
Flat metal, jacketed asbestos-filled:				
Soft aluminum	3.25	5 500		(1a), (1b), (1c),‡ (1d),‡ (2)‡; column II
Soft copper or brass	3.50	6 500		
Iron or soft steel	3.75	7 600		
Monel or 4-6% chrome	3.50	8 000		
	3.75	9 000		
Stainless steels	3.75	9 000		

TABLE 3.2 Gasket Materials and Contact Facings (*Cont.*)

Gasket Materials and Contact Facings†
Gasket Factors m for Operating Conditions and Minimum Design Seating Stress y

Gasket material	Gasket factor m	Minimum design seating stress y, psi	Sketches	Facing sketch and column to be used from Table 26-4
Self-energizing types (O-rings, metallic, elastomer, other gasket types considered as self-sealing)	0	0		
Elastomers without fabric or high percentage of asbestos fiber:				(1a), (1b), (1c), (1d), (4), (5); column II
Below 75A Shore Durometer	0.50	0		
75A or higher Shore Durometer	1.00	200		
Asbestos with suitable binder for operating conditions:				(1a), (1b), (1c), (1d), (4), (5); column II
⅛ in thick	2.00	1 600		
1/16 in thick	2.75	3 700		
1/32 in thick	3.50	6 500		

Gasket Materials and Contact Facings†
Gasket Factors m for Operating Conditions and Minimum Design Seating Stress y

Grooved metal:				(1a), (1b), (1c), (1d), (2), (3); column II
Soft aluminum	3.25	5 500		
Soft copper or brass	3.50	6 500		
Iron or soft steel	3.75	7 600		
Monel or 4–6% chrome	3.75	9 000		
Stainless steels	4.25	10 100		
Solid flat metal:				(1a), (1b), (1c), (1d), (2), (3), (4), (5); column I
Soft aluminum	4.00	8 800		
Soft copper or brass	4.75	13 000		
Iron or soft steel	5.50	18 000		
Monel or 4–6% chrome	6.00	21 800		
Stainless steels	6.50	26 000		
Ring joint:				(6); column I
Iron or soft steel	5.50	18 000		
Monel or 4–6% chrome	6.00	21 800		
Stainless steels	6.50	26 000		

†This table gives a list of many commonly used gasket materials and contact facings with suggested design values of m and y that have generally proved satisfactory in actual service when using effective gasket seating width b given in Table 26-4. The design values and other details given in this table are only suggested and are not mandatory.
‡The surface of a gasket having a lap should not be against the nubbin.

TABLE 3.3 Effective Gasket Width

Gasket Materials and Contact Facings†
Gasket Factors m for Operating Conditions and Minimum Design Seating Stress y

Gasket material	m	Min. design seating stress	Facing sketch	Use facing sketch
Grooved metal:				
Soft aluminum	3.25	5 500		(1a), (1b), (1c), (1d), (2), (3); column II
Soft copper or brass	3.50	6 500		
Iron or soft steel	3.75	7 600		
Monel or 4–6% chrome	3.75	9 000		
Stainless steels	4.25	10 100		
Solid flat metal:				
Soft aluminum	4.00	8 800		(1a), (1b), (1c), (1d), (2), (3), (4), (5); column I
Soft copper or brass	4.75	13 000		
Iron or soft steel	5.50	18 000		
Monel or 4–6% chrome	6.00	21 800		
Stainless steels	6.50	26 000		
Ring joint:				
Iron or soft steel	5.50	18 000		(6); column I
Monel or 4–6% chrome	6.00	21 800		
Stainless steels	6.50	26 000		

†This table gives a list of many commonly used gasket materials and contact facings with suggested design values of m and y that have generally proved satisfactory in actual service when using effective gasket seating width b given in Table 26-4. The design values and other details given in this table are only suggested and are not mandatory.
‡The surface of a gasket having a lap should not be against the nubbin.

Simplified procedure

A simpler method of calculation has been suggested by Whalen. This method is also based on the seating stress S_g on the gasket, as shown in Table 3.4, and on the hydrostatic end force involved in the application. Basically, Whalen's equations accomplish the same thing as the code, but they are simplified since they use the full gasket contact width, regardless of the flange width and the surface finish of the sealing faces.

This method is based on the total bolt load F_b being sufficient to:

1. Seat the gasket material into the flange surface.
2. Prevent the hydrostatic end force from unseating the gasket to the point of leakage.

In the first case, Table 3.4 lists a range of minimum seating stress values. The ranges shown were found in a search of the literature on gasket seating stresses. Gasket suppliers can be contacted to confirm these values.

As a starting point in the design procedure, the mean value of S_g could be used. Then, depending on the severity of the

TABLE 3.4 Minimum Recommended Seating Stresses for Various Gasket Materials

Material	Gasket type	Minimum seating stress range S_g, psi*
Nonmetallic:		
Asbestos fiber sheet	Flat	
$\frac{1}{8}$ in thick		1400–1600
$\frac{1}{16}$ in thick		3500–3700
$\frac{1}{32}$ in thick		6000–6500
Asbestos fiber sheet $\frac{1}{32}$ in thick	Flat with rubber beads	1000–1500 lb/in on beads
Asbestos fiber sheet $\frac{1}{32}$ in thick	Flat with metal grommet	3000–4000 lb/in on grommets
Asbestos fiber sheet $\frac{1}{32}$ in thick	Flat with metal grommet and metal wire	2000–3000 lb/in on wire
Cellulose fiber sheet	Flat	750–1100
Cork composition	Flat	400–500
Jacketed metal-asbestos:		
Aluminum	Plain	2,500
Copper	Plain	4,000
Carbon steel	Plain	6,000
Stainless steel	Plain	10,000
Aluminum	Corrugated	2,000
Copper	Corrugated	2,500
Carbon steel	Corrugated	3,000
Stainless steel	Corrugated	4,000
Stainless steel	Spiral-wound	3,000–30,000
Metallic:		
Aluminum	Flat	10,000–20,000
Copper	Flat	15,000–45,000 depending on hardness
Carbon steel	Flat	30,000–70,000 depending on alloy and hardness
Stainless steel	Flat	35,000–95,000 depending on alloy and hardness
Aluminum (soft)	Corrugated	1000–3700
Copper (soft)	Corrugated	2500–4500
Carbon steel (soft)	Corrugated	3500–5500

TABLE 3.4 Minimum Recommended Seating Stresses for Various Gasket Materials (*Cont.*)

Material	Gasket type	Minimum seating stress range S_g, psi*
Stainless steel	Corrugated	6000–8000
Aluminum	Profile	25,000
Copper	Profile	35,000
Carbon steel	Profile	55,000
Stainless steel	Profile	75,000

*Stresses in pounds per square inch except where otherwise noted.

application and/or the safety factor desired, the upper and lower figures could be utilized.

Two equations are associated with this procedure. The first is

$$F_b = S_g A_g \tag{3.6}$$

where F_b = total bolt load, lb
S_g = gasket seating stress, psi (from Table 3.4)
A_g = gasket contact area, in²

This equation states that the total bolt load must be sufficient to seat the gasket when the hydrostatic end force is not a major factor. The second equation associated with the hydrostatic end force is

$$F_b = KP_1 A_m \tag{3.7}$$

where P_1 = test pressure or internal pressure if no test pressure is used
A_m = hydrostatic area on which internal pressure acts (normally based on gasket's middiameter)
K = safety factor

The safety factors K from Table 3.5 are based on the joint conditions and operating conditions but not on the gasket type or flange surface finish. They are similar to the m factors in the ASME code. The equation using K states that the total bolt load must be more than enough to overcome the hydrostatic

Chapter Three

TABLE 3.5 Safety Factors for Gasketed Joints

K factor	When to apply
1.2–1.4	For minimum-weight applications where all installation factors (bolt lubrication, tension, parallel seating, etc.) are carefully controlled; ambient to 250°F temperature applications; where adequate proof pressure is applied
1.5–2.5	For most normal designs where weight is not a major factor, vibration is moderate, and temperatures do not exceed 750°F. Use high end of range where bolts are not lubricated
2.6–4.0	For cases of extreme fluctuations in pressure, temperature, or vibration; where no test pressure is applied; or where uniform bolt tension is difficult to ensure

end force. The middiameter is used in A_m since testing has shown that just prior to leakage, the internal pressure acts up to the middiameter of the gasket.

After the desired gasket has been selected, the minimum seating stress, as given in Table 3.4, is used to calculate the total bolt load required by multiplying the seating stress and the gasket contact area [Eq. (3.5)]. Then the bolt load required to ensure that the hydrostatic end force does not unseat the gasket is calculated from Eq. (3.7). The total bolt load F_b calculated by Eq. (3.5) must be greater than the bolt load calculated in Eq. (3.7). If it is not, the gasket design must be changed, the gasket's area must be reduced, or the total bolt load must be increased.

Gasket selection

ASME code. Note: These are annular gaskets.

N = number of bolts
T = bolt torque, lb-in
D = bolt diameter, in
K = nut factor, e.g., 0.16 or 0.020 depending on lubrication
P = maximum working pressure, psi
G = gasket load-bearing diameter, in
b = gasket load bearing width, in
F_a = assembly bolt force, lb

$$F_a = \frac{TN}{KD}$$

W_{m_2} = ASME gasket seating load

$$W_{m_2} = 3.14bGy$$

For gasket to seat: $F_a \geq W_{m_2}$

$$F_a \geq 3.14\, bGy$$

$$by \leq \frac{F_a}{3.14\, G}$$

$$by \leq \frac{TN}{3.14\, KDG}$$

For gasket to seal during operation, $F_a \geq W_{m_1}$

$$F_a \geq \frac{(3.14)G^2P}{4} + 2b(3.14)\, G_{mp}$$

$$bm \leq \frac{F_a - 3.14G^2P/4}{2(3.14)GP}$$

$$bm \leq \frac{(TN/RD)-(\pi/4)G^2P}{2\pi GP} \tag{3.8}$$

The designer selects the gasket width and its m factor to comply with Eq. (3.8).

ASME code (general sealing)

A_p = pressure area, in^2

A_g = gasket seating area, in^2

$$F_a = \frac{TN}{KD}$$

$$W_{m_2} = A_gY$$

For gasket to seal:

$$F_a \geq W_{m_2}$$

$$AgY \leq \frac{TN}{KD}$$

$$F_a \geq W_{m_1}$$

$$F_a \geq A_p P + 2mA_g P$$

$$mA_g P \leq F_a - A_p P$$

$$m \leq \frac{F_a - A_p P}{Ag\, P}$$

$$m \leq \frac{(TN/KD) - A_p P}{Ag\, P}$$

or

$$m \leq \frac{F_a - F_s}{F}$$

where F_a = assembly bolt force
F_s = flange separation force
F = minimum gasket contact force

Simplified process

$$F_b = \text{total bolt load, lb}$$
$$A_m = \text{hydrostatic area, in}^2$$
$$P_i = \text{proof test pressure, psi}$$

Calculate factor of safety

$$K = \frac{F_b}{P_i A_m}$$

If k is high enough, proceed. Otherwise change joint design (go back to A). Use F_b to select gasket based on seating stress S_g and contact area A_g,

$$S_g A_g \leq F_b$$

The designer should verify that gasket crush does not occur during assembly or operation.

Potential New Gasket Code

As noted earlier, the ASME code has been used by industry for more than 50 years. The factors were introduced by Ross, Heim, and Markl and described in the *Mechanical Engineering* magazine article "Gasket Loading Constants," written in September 1943.

The factors have served industry well, but many developments over the years in gaskets and fasteners imply that their validity today should be verified. During the last 10 years the Pressure Vessel Research Committee (PVRC) has conducted a gasket program to prove and improve the code's design factors. At this time the basic testing has been completed and a special working group (SWG) of ASME is studying the PVRC findings. There are two basic changes that most likely will happen:

1. A new appendix which parallels Appendix 2 of Sec. VIII, Div. 1 of the ASME code will be issued. This appendix will contain tables with new gasket constants. These constants are identified as G_b, a, and G_s. These constant are defined later in this chapter.

2. A new Standard Test Method utilizing the new gasket constants for various gaskets will be issued.

Gasket testing

The PVRC testing commenced with the determination of a gasket's load-deformation characteristics expressed in terms of stress on the gasket versus gasket deflection. Figure 3.11 depicts a typical curve.

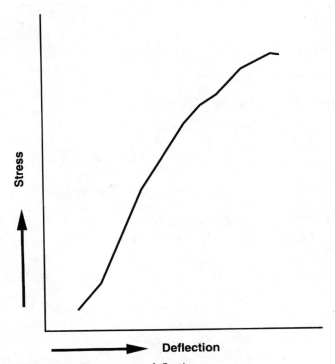

Figure 3.11 Stress versus deflection.

After loading, the gasket was unloaded and the unload stress versus deflection was determined. Figure 3.12 shows a typical graph.

During these load and unload conditions, the gasket was pressurized for leakage detection. This was done at a number of initial stress levels. Figure 3.13 depicts these data at two leak rates A and B.

The investigators found that plotting the log of the test pressure versus the log of the mass leak rate at various stress levels resulted in linear relationships and the determination of a dimensionless parameter called the tightness parameter, T_P (see Fig. 3.14).

A tightness parameter of 100 would mean a test pressure of 100 atmospheres (1470 psi). A total leak rate of 1 mg/s would occur from a gasket having a 150-mm (6-in) diameter. The tightness parameter relationship is as follows:

The Gasket and the Joint 119

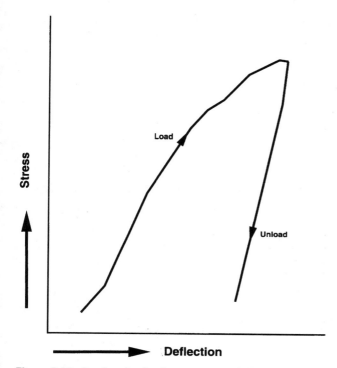

Figure 3.12 Load and unload stress versus deflection.

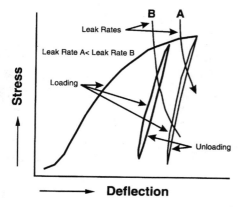

Figure 3.13 Stress versus deflection and leak rates.

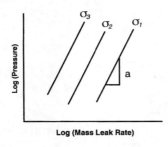

$$T_p = \frac{KP}{L_m^a}$$

Figure 3.14 Pressure versus leak rate.

$$T_p = \frac{P}{P^*} \times \left(\frac{L^*_{RM}}{L_{rm} D_t}\right)^a$$

where T_p = tightness parameter (dimensionless)
 P = contained pressure, psi or kPa
 P^* = reference atmospheric pressure (14.7 psi or 1/1.3 kPa)
 L^*_{RM} = standard leak rate (lb/h-in, mg/s-mm) which is keyed to a normalized reference gasket of 5.9 in (150 mm) outside diameter
 a = experimentally determined exponent (e.g., 0.5 if the contained fluid is a gas, 1.0 if it's a liquid)
 D_t = gasket OD, in, mm
(And see note below.)*

Volumetric leak rates are more common than are mass leak rates. Table 3.6 shows some appropriate volumetric leak rates for the referenced mass leak rate of (0.008 lb/h-in or 1 mg/s-mm.

Figure 3.15 is a loglog plot of gasket stress versus tightness parameter for a gasket. It contains both the loading and unloading conditions. This type of plot is the basis for the PVRC gasket constants. Another important PVRC decision concerns tightness. Three tightness classifications have been selected. Table 3.7 depicts the tightness callout, the classification, and the mass leak rates associated with the three classifications.

---

*Note: If T_p = 10, an internal pressure of 147 psi is needed to create an L_{rm} leak rate of 1 mg/s of liquid from a 150-mm OD gasket.

TABLE 3.6 Volumetric Equivalence for Mass Leak Rates

Fluid	Approximate Volumetric Leak Rate
Nitrogen -	7 pts/hr or 1 ml/sec
Helium -	50 pts/hr or 6 ml/sec
Water -	0.01 pts/hr or 1 x 10 ml/sec

Leak Rate and Volumetric Equivalent	
Leak Rate (ml/sec)	Volumetric Equivalent
10^{-1}	1 ml/10 sec
10^{-2}	1 ml/100 sec
10^{-3}	1 ml/1,000 sec
10^{-4}	1 ml/10,000 sec

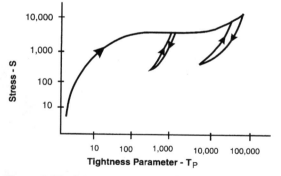

Figure 3.15   Stress versus tightness parameter.

TABLE 3.7 Proposed Tightness Classes

Tightness classification	Classification	Mass leak rate per mm of diameter
Economy	T1	1/5 mg/sec.mm
Standard	T2	1/500 mg/sec.mm
Tight	T3	1/50,000 mg/sec.mm

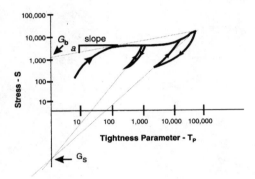

Figure 3.16 Stress versus tightness parameter in identified gasket factors or constants.

As noted earlier, the PVRC gasket factors or constants are G_b, a, and G_s. These factors are identified in Fig. 3.16.

G_b is the upper intercept of the gasket stress axis. This is associated with the loading parameter of the test.

a is the slope of the loading portion of the test.

G_s is the lower intercept of the gasket stress axis. This is associated with the unloading portion of the test. Note that this intercept is associated with $T_p = 1$. Some of the PVRC data show this intercept at $T_p = 10$.

G_b and a are constants associated with the seating load on the gasket. This is like the old y factor. G_b and a will determine the initial sealing stress on a gasket for a selected

TABLE 3.8 Values of Gasket Tightness Constants for Various Gaskets

Type of Gasket	Tightness Factors		
	G_b psi	a	G_s psi
Premium Compressed asbestos with elastomer binder (1/16 thick)	2500	0.15	117
Premium Compressed asbestos with elastomer binder (1/8 thick)	400	0.38	15
Premium glass fiber with elastomer binder (1/16 thick)	1150	0.30	117
Premium glass fiber with elastomer binder (1/32 thick)	285	0.38	117
Laminated flexible graphite with chemicaly bonded stainless steel insert	450	0.45	1E-4
St. steel, double jacketed with asbestos millboard or mica filler: flat (1/8 thick)	2900	0.23	15
Flat, low-carbon steel, corrugated piping gasket with mica filler (1/8 thick)	3400	0.25	200
Spiral wound, Class 600, with st. steel windings and flexible graphite filler	1300	0.30	15

tightness level. G_s is related to the old maintenance factor, m, since it is associated with the unloading of the gasket.

Table 3.8 lists some values of G_b, a, and G_s for various gaskets.

Gasket selection

The following is the design procedure based on the PVRC tightness criterion:

1. Establish minimum tightness.

$$T_{p_{\min}} = 1.8257\, c \left[\frac{P}{P^*}\right]$$

where c = 0.1 for T_1—economy tightness
      = 1.0 for T_2—standard tightness
      = 10 for T_3—tight tightness

2. Determine seating stress.

$$S_{y_a} = \frac{G_b}{e} (1.5\, T_{p_{\min}})^a$$

where $e = 0.75$ for manual bolt-up
$ = 1.0$ for ideal bolt-up

See Fig. 3.17.

3. Determine minimum "design" stress S_m. S_m is the greater of S_{m_1}, S_{m_2}, or $2p$. For seating:

$$S_{m_2} = \frac{S_{y_a}}{1.5} - P \frac{A_1}{A_g}$$

For operation:

$$S_{m_2} = G_s \left(e\, \frac{S_{y_a}}{G_s} \right)^{\frac{1}{T_r}}$$

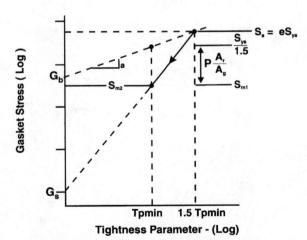

Figure 3.17 Stress versus tightness and seating identification.

where $T_r = \log(1.5T_{p_{\min}})\backslash\log(T_{p_{\min}})$

Here $1.5 T_p$ min is the assembly tightness and $T_{p_{\min}}$ is the operating tightness

NOTE: The 1.5 factor that reduces S_{ya} gives credit for the fact that bolts are normally tightened to 1.5 (or more) times their design allowable during assembly of the joint. Such tightening is consistent with the hydrostatic test.

4. Calculate the design bolt load

$$W_m = PA_i + S_m A_g$$

where A_i = pressurized area
A_g = gasket area

References

Bazergui, A., and L. Marchand: "Development of a Production Test Procedure for Gaskets," *Welding Research Council Bulletin* 309, November 1985.
Code for Pressure Vessels, American Society of Mechanical Engineers, Sec. VIII, Div. 1, App. 2, 1980.
Czernik, D. E.: "Recent Developments and New Approaches in Mechanical and Chemical Gasketing," SAE paper 810367, February 1981.
Czernik, D. E., J. C. Moerk, Jr., and F. A. Robbins: "The Relationship of a Gasket's Physical Properties to the Sealing Phenomena," SAE paper 650431, May 1965.
Payne, J. R., A. Bazergui, and G. F. Leon: "Getting New Gasket Design Constants from Gasket Tightness Data," Special Supplement, *Experimental Techniques,* November 1988, pp. 22–27.
Rothbart, H. A.: *Mechanical Design and Systems Handbook,* 2d ed., McGraw-Hill, New York, 1985, sec. 27.4.
Standard Handbook of Machine Design, McGraw-Hill, New York, 1986, chap. 26, part 1.
Whalen, J. J.: "How to Select the Right Gasket Material," *Product Engineering,* October 1960.

Chapter 4

Pressure Vessel Research Council

PVRC Gasket Testing and Analysis

Preventing excessive liquid or gas leaks is one of the most important and most difficult jobs gaskets have to accomplish during their service life under various conditions. Recently, this task has become even more difficult to achieve owing to the use of asbestos-substitute fibers and the required reduction of emissions imposed by an increasing environmental consciousness. It is more important than ever to avoid large and small leaks.

In North America, a substantial research effort has been dedicated over the last decade to the study of flanged joints and the characterization of gasket behavior at room and elevated temperatures. Through the auspices of the Pressure Vessel Research Council (PVRC), the Materials Technology Institute of Chemical Process Industries (MTI), and others, a fundamental understanding of gasket behavior has been achieved in parallel with the development of new gasket test methods.

Some of the research work has matured to the point where it has become suitable for incorporation in the ASME Pressure Vessel Code (Sec. VIII, Div. 1) in the form of new gasket constants and an alternative bolt load calculation procedure for designs that are based on flanged joint tightness.[22] At the same time, the room-temperature tightness test and many of the new elevated-temperature gasket test methods are in the process of being adopted as ASTM standard tests.

This chapter summarizes some of the work performed over the past 10 years in the field of bolted flange joints, gives an up-to-date status (as per 1994) of the research fallout, and presents an overview of the continuing research effort.

PVRC Gasket and Bolted Joint Research Programs

As a consequence of questions about the origin and replication of gasket factors, the PVRC Committee on Bolted Flanged Connections was asked by the ASME to conduct gasket research. In 1974, a room-temperature exploratory test program was initiated with committee guidance. The details and results of the exploratory program are documented in references 7 through 10.

A major finding of the exploratory program was that the current ASME code gasket factors must be modified to show dependence of tightness, contained medium, assembly stress, and internal pressure. This finding led to Gasket Test Program II. It was anticipated that the expanded research would provide a better understanding of the sealing mechanism and develop more meaningful gasket design factors. Another goal was to develop a standard test procedure for gasket constants based on tightness.

The availability of new gasket materials and styles whose properties are different from those of traditional asbestos products has made achieving these goals even more relevant.

Three major room-temperature gasket test programs are associated with the PVRC effort since the initial ASME request. They are:

- Exploratory Gasket Test Program (1974–1983 Room Temperature)
- Gasket Test Program II (1979–1989 Room Temperature—Milestone, Development, and Production Tests)
- Elevated Temperature Gasket Testing and Bolted Flange Joint Behavior Program

These programs are summarized by Leon, Bickford, Hsu, Winter, and Payne[11,12,13] and in greater detail by various *Welding Research Council Bulletins.*

Gasket Test Program II—Room Temperature

This program began in 1979. It was formulated by the PVRC in cooperation with cognizant committees of the ASME, the American Petroleum Institute, ANSI, the ASTM, and the Fluid Sealing Association. Funding outside normal PVRC budgets is by 32 sponsors from five countries.

The major technical areas of development for Program II consisted of:

1. Gasket leakage test literature study, completed 1979
2. Flanged joint user experience, completed 1983
3. Milestone tests, completed 1983
4. Production test procedure development, completed 1985

The gasket mechanical and leakage performance data for over three dozen different gasket styles and materials have been obtained.

Elevated-temperature test and analytical programs

A multiphase PVRC elevated-temperature program was initiated in 1982. The program consists of both analytical and experimental efforts, and its overall objectives are to:

- Gain a fundamental understanding of bolted joint behavior and mechanisms which lead to leakage of such joints.
- Determine the relative contribution of all factors affecting leakage.
- Develop design procedures to minimize bolted joint leakage.

The elevated-temperature program is funded outside normal PVRC budgets by 14 sponsors from four countries. References 12 through 14 track progress of the program.

It should be noted that the PVRC elevated-temperature gasket test effort has paralleled and supplemented the MTI Project 47 elevated-temperature gasket evaluation effort.[15,16] These two programs have developed standardized test procedures that enable manufacturers and users to screen and qualify gaskets for high-temperature service. Between them the following test procedures were developed:

- Fire Simulation Screen Test (FIRS)
- Aged Tensile and Relaxation Screen Test (ATRS and HATR)
- Hot Operational Tightness Tests (HOTT and AHOT)
- Aged Relaxation Leakage Adhesion Screen Test (ARLA)
- Fire Simulation Tightness Test (FITT)

Committee plans

Current plans of the PVRC Committee on Bolted Flanged Connections are focused on the following areas:

Gasket testing and methods. Continue gasket testing, supplementing existing data for other media and long-term exposure; develop improved test and qualification schemes, fugitive emissions characteristics, and fire evaluation.

Flange rating. Develop tools and criteria for rating standard flanged joints by establishing the effect and importance of parameters such as tightness, thermal transients, relaxation, rotation, and external loads.

Flange design. Develop improved design criteria and modeling for all types of ASME joints considering parameters such as tightness, geometry, facing, thermal transients and cycles, relaxation, rotation, external loads, and assembly.

Flanged joint assembly and interaction effects. Continue development of improved and more rapid assembly methods; develop guidelines for qualifying bolt-up procedures and personnel.

Implement developments. Continue ASME SWG and ASTM gasket (Committee F3) activities to implement PVRC constants and test method developments.

Design guidelines report. Prepare report on current design and analysis guidelines for bolted joints. Consider items such as modeling, tightness, new constants, transients, relaxation, rotation, and external loads. Identify data and development needs.

Room-temperature research program, a revision of the ASME Code *m* and *y* gasket factors

Because the mechanical characteristics and the sealing behavior of a gasket are complex and specific data are generally unavailable, the design of gasketed joints has been reduced to a series of simplified equations based on two experimental "gasket factors," respectively, the *y* factor for gasket seating yield and the *m* factor for load maintenance in pressurized service. The rules of the ASME Boiler and Pressure Vessel Code[1] and these gasket factors introduced by Rossheim and Markl in 1943[17] have been widely used in the design of gasketed bolted flanged joints. This procedure has been successful, at least in the pressure vessel world, and is still in use. Generally the *m* and *y* factors have served industry well. One shortcoming has been materials because there is no workable standard test procedure. Further, in the ASME code, the gasket constants were presented as "suggestions" for the designer and were never intended to define leak-free gasketed joints in service. In these environmentally sensitive times, there is a need for an approach to bolted joint design that considers leakage and makes the tightness of a joint a design criterion. Therefore, confirmation and improvement of such design factors was more than necessary.

At the request of ASME, the Pressure Vessel Research Committee (PVRC) directed the sponsor-funded room-temperature gasket test program, which has the following goals:

- Better understand the mechanical characteristic of gaskets
- Better understand the sealing mechanism
- Develop more meaningful gasket design factors
- Develop a standard test procedure at room temperature

- Develop design procedures to minimize leakage of gasketed flanged joints

Mechanical Behavior of Gaskets

The gasket can be considered as a spring in series with bolts, nuts, washers, and flange springs. Since the gasket is part of the joint, we can combine its spring constant with that of the flange members to construct a joint diagram and make joint calculations. However, the gaskets have usually very unpleasant characteristics such as:

- The stiffness of the gasket being usually smaller than that of the joint members, it dominates the elastic behavior of the joint.
- The stiffness of the gasket is usually nonlinear during its initial compression but becomes more linear for subsequent unloading and reloading. The gasket exhibits a great deal of hysteresis and will take a permanent set.
- The gasket has a tendency to creep under load. This reduces the tension in the bolts, often before the joint has been pressurized. Creep can be substantially increased by elevated temperatures.

Test data on the mechanical behavior of various types of gasket materials were scarce until some work was initiated by the PVRC in 1983–1984. Three types of test at room temperature were proposed involving:

- Low cycle loading-unloading behavior
- Creep under constant gasket stress
- Gasket stress relaxation at constant gasket deflection

None of these tests represents the true bolted joint situation, which would involve simultaneous creep and relaxation, but these tests can be used to estimate the amount of creep and relaxation which might be encountered in practice. Such tests are now combined in one test procedure called the Room Temperature Mechanical Test (ROMT). This is a test procedure utilizing a hydraulic fixture to evaluate the mechanical rigidity and creep behavior of gaskets at room temperature. The gasket is loaded at room temperature and load is then cycled three times between low and high stress levels at a rate of 100 psi/s. At the end of the third load cycle, the gasket is allowed to creep at constant load or to relax at constant gasket deflection for 5 h. An elevated-temperature version of this test, called the Hot Mechanical Test (HOMT), has also been developed.

Elevated-Temperature Research Program

The long-term goal of the elevated-temperature research program is to develop design procedures that will minimize leakage of gasketed bolted flanged joints in elevated-temperature service. Although this ultimate goal is not yet entirely reached, research work accomplished so far has led to basic understanding of gasket behavior at elevated temperature and the mechanisms which can lead to leakage of gasketed bolted flanged joints. This work has also resulted in many new gasket testing methods suitable for future incorporation in world class standards. Also, qualification guides which serve to identify the pertinent characteristics and the probable long-term performance of different types of gasket for chemical and petrochemical plant service have been proposed.

Gasket behavior at elevated temperature

The proper assessment of gasket performance requires that the hot mechanical and tightness properties of the gasket material be known for elevated-temperature process industry application to bolted joints. This is especially true for gasket materials where time-dependent temperature effect plays an important role by inducing thermal degradation of gasket components. Aging is defined as the effect of the thermal degradation, over time and temperature, on the physical properties of gasket materials under specific working conditions (gasket compressive stress, internal fluid type and pressure, gasket type and geometry, flange rigidity). For gaskets in elevated-temperature service, mechanical stability, creep and relaxation resistance, and weight loss of gasket materials are considered mechanical key properties that should be monitored with time to determine whether a gasket will maintain a safe long-term tightness performance of a joint. This is because it is suspected that strong correlation exists between hot mechanical properties ensuring structural stability of a gasket and its leakage behavior during aging. Therefore, proper assessment of gasket performance at elevated temperature requires that these key properties be known over a period of time consistent with service temperature expectations of the different gasket styles.

In general, gasket products can be tentatively classified in four different groups or classes depending on their elevated-temperature behavior characteristics.

Elastomer bound sheet materials

Gasket properties such as load relaxation, creep, residual tensile strength, and tightness will experience gradual degradation as a function of temperature and exposure time.[15] Decomposition and oxidization of a material's organic components (binder, organic fibers) are the main cause of degradation. Property changes are strongly related to gasket weight loss during thermal exposure.[25] Apart from temperature and time, some important parameters have an influence on the degradation rate, namely, internal fluid and pressure, gasket geometry, initial compression load, and flange rigidity.[26]

PTFE based sheets, ribbons, etc.

Load relaxation resistance and creep behavior of these materials relate directly to temperature, and exposure time has little if no effect. They are known to be good sealers, but when unloaded, PTFE based gaskets could be susceptible to blowout.

Flexible graphite sheets

It is well recognized that metal-reinforced flexible graphite sheet gaskets are prime candidates for nonasbestos elevated-temperature applications. At temperatures above 750°F, these gaskets exhibit increased degradation with time owing to oxidization reactions with surrounding air. Test results also indicate that under low compression stress, the degradation process is substantially accelerated. A significant acceleration of the oxidation has also been observed when a metallic reinforcement is inserted between the flexible graphite sheets to increase the mechanical resistance. When these materials are well protected against oxidation in the flange or used with the addition of a passivating corrosion and oxidation inhibitor, they can sustain much higher service temperature.

Composite gaskets

Spiral-wound gaskets, double-jacketed gaskets, etc., containing filler materials (i.e., organic, PTFE, flexible graphite binders) are susceptible to thermal degradation and will exhibit a significant change in tightness behavior resulting in increased leakage and/or tightness sensitivity to stress excursion.[27]

Quantification of Aging Effects

In order to quantify and compare the aging effect on the hot performance of gasket products, whether in sheet form or of composite construction, it has become necessary to develop:

- New reliable hot test methods because existing standardized methods for gaskets are unsatisfactory (ASTM F36, F37, F38, and also DIN 52913, 3535, etc.).
- New reliable quantification tools that recognize aging as a fundamental characteristic of gaskets in elevated-temperature service. Such tools should quantify the combined effect of temperatures and exposure times on some physical gasket properties that are considered as key factors to characterize the degradation due to aging.
- New realistic criteria of acceptance to determine the acceptability of new gaskets or gasket materials and make comparison with well-known gasket products.
- New reliable qualification test protocols for each gasket class or gasket styles to determine the realistic long-term service conditions for a broad range of applications (chemical and petrochemical services).

The Pressure Vessel Research Council (PVRC) and the Materials Technology Institute of the Chemical Process Industries (MTI) have sponsored much developmental work. One of the objectives of MTI Project 47 was to develop and assess a test procedure of evaluation of long-term hot mechanical properties of asbestos replacement materials subjected to the thermal degradation of an oxidizing medium like air. This procedure, known as the Aged Tensile Relaxation Screen Test (ATRS), is presented below. The ATRS screen test data are obtained in still air environment. As a first simple attempt to reproduce the aging effect on gasket properties, the air exposure was a valid approach because this environment represents the worst condition encountered by users in process plants and because most gaskets in process applications have one edge air exposed. We believe steam and liquids are less harsh unless chemical attack is involved.

- HOMT (Hot Mechanical Test). Test procedure utilizing a hydraulic fixture to evaluate the mechanical rigidity and creep behavior of gaskets at elevated temperature. The gasket is loaded at room temperature and heated up to the required temperature. Load is then cycled three times, between low and high stress levels, at a rate of 100 psi/s. At the end of the third cycle, the gasket is allowed to creep, at constant load, for 5 h.

Fire resistance tests

- FIRS (Fire Resistance Screen Test). Developed as a preferred alternative to open-flame test methods. Uses springless ATRS fixture to expose gasket specimens to simulated fire conditions (1200°F, 15 min) in a still air oven. Gasket appearance, residual tensile strength, and load relaxation permit a judgment on the ability of a gasket to survive fire exposure.
- FITT (Fire Tightness Test).[30] A fire survival test carried out on a 4-in NPS gasket. During the test, gasket internal helium pressure is maintained constant at 400 psi while its tightness is measured under simulated fire conditions (1200°F for 15 min).

Elevated-temperature tightness tests

- HOTT (Hot Operational Tightness Test).[31] Extends the PVRC room-temperature tightness test (ROTT) concept to elevated temperatures up to 1200°F. The HOTT test realistically reproduces pressures, bolt loads, and temperatures used in process and power plants. The test procedure involves initial compression at room temperature (simulation of conditions, and stress excursions and thermal disturbance cycles).

- AHOTT (Aged Hot Operational Tightness Test).[32] An alternative to the HOTT procedure for situations requiring long aging periods (more than 10 days). It differs from the HOTT procedure only during the aging period whereby the precompressed gasket and platen assembly is aged in a still air oven instead of the HOTT fixture. A low internal gasket pressure (air, nitrogen, or helium) can be used to analyze the gas effect on the gasket degradation.

- The HOTT and AHOT are performed on the Universal Gasket Test Rig (UGR).

Standardization of elevated-temperature tests. Drafts of ASTM standards for the ATRS, FIRS, HOTT, and AHOT tests have been under study by the F3 Committee on Gasket Testing. The drafts are presently in a revision process and should be submitted for initial approval in the near future. The ASTM will then proceed with final round-robin testing. All these tests will be used as screening tests. Results will be interpreted in accordance with future specification schemes.

New Quantification Tools

These tools serve two purposes:

- Provide a common basis for comparing the damage due to the combined effect of temperature and time of one set of test conditions to that inflicted by another set of time and temperature conditions and therefore are very useful for setting the test conditions for follow-up tests.

- Provide a first rational basis to predict long-term performance of a sheet gasket material from short-term hot test data.

Quality parameters

Depending on which key properties are used to express the gasket damage due to temperature on various gasket styles (gasket stress retention capability, residual tensile strength, weight loss of gasket materials, etc.), several quantification tools have been developed. Such tools, called quality parameters, were primarily derived from elevated-temperature screening tests (ATRS/HATR) performed on numerous types of elastomer bound sheet gaskets, but their use has been extended to other types of materials.

Mechanical quality parameters Q_r, Q_{tx}. Two mechanical properties were considered initially as critical aspects of gasket performance as they affect the integrity of a gasketed bolted joint.

Load retention measures how much the gasket contributes to the reduction of bolt load. Tensile strength tells us if the gasket will "fall apart" in the event stud load is reduced to the point of near blowout conditions of low gasket stress in the joint. These two properties were selected because they represent a primary gasket mechanical quality basis that gives an assurance of freedom from catastrophic failure in the form of gross leaks in operating bolted joints.

It is clear that the postaging tensile strength is not always representative of the thermal degradation of all gasket materials or gasket styles. For unreinforced elastomeric sheet materials, tensile strength is certainly a good representation of the material integrity while for metal-reinforced products it is not. On the other hand, the use of residual tensile strength presents a question of interpretation which was not completely addressed in the MTI project, that is, whether postexposure tensile strength above some threshold value is meaningful for the tightness performance of gaskets. Relaxation is more important than tensile strength once it is established that the material has adequate strength so that it will not "fall apart" and has sufficient integrity that it will not leak too much.

Since there is a large body of experience with asbestos fiber reinforced sheet materials, its use as a quality baseline is appropriate. The threshold values proposed for reference in the definition of quality parameters indicate that asbestos materials do not "fall apart" and maintain an acceptable load retention, at least after the above exposures experienced during the MTI project. Postaging test properties for much longer exposures (time and temperature) have been measured and the results confirmed that these materials are still acceptable for operating temperatures to 750°F or more.

A dimensionless measure of load retention quality Q_r based on percent stud bolt load retained is defined as the square of the ratio

$$Q_r = \frac{\text{percent retained}}{75} \times 2$$

- Percent retained is the percentage of compressive load retained on the fixture by three layers of aged specimens installed in the ATRS fixture.
- A 75 percent stud bolt load retained is proposed as a reference measure of quality based on test results obtained from short-term ATRS test data (700°F maximum and 42 days exposure) performed on asbestos materials.
- The use of a squared function for the load retained acknowledges the importance of the load relaxation effect in gross failures of hot bolted joints in process plants.

A dimensionless measure of tensile quality Q_{tx} based on the posttensile strength of ATRS gasket samples is defined as the ratio:

$$Q_{tx} = \frac{TSX}{1000} \quad \text{if } TSX < 2500 \text{ psi}$$

$$Q_{tx} = 2.5 \quad \text{if } TSX \geq 2500 \text{ psi}$$

- TSX is the remaining tensile breaking strength (psi, MPa) in the weak direction of ATRS aged gasket coupons.
- A residual tensile strength of 1000 psi (6.9 MPa) is proposed as a reference measure of tensile quality based on test

results obtained from short-term ATRS test data (700°F maximum and 42 days' exposure) performed on asbestos materials.
- An interpretation that has been applied in regard to the usefulness of high tensile strengths is to give no credit to tensile strength exceeding 2500 psi.

An overall mechanical quality parameter (developed for the MTI). A convenient way to express the interaction between the two mechanical key properties as they affect the performance of elastomeric fiber sheet materials was to combine them in a single relationship and define a dimensionless quality parameter Q_p as follows:

$$Q_p = Q_{tx} \times Q_r$$

$$Q_p = \frac{TSX}{1000} \times \left(\frac{\text{percent retained}}{75}\right)^2 \quad \text{if } TSX < 2500 \text{ psi}$$

$$Q_p = 2.5 \times \left(\frac{\text{percent retained}}{75}\right)^2 \quad \text{if } TSX \geq 2500 \text{ psi}$$

The quality parameter Q_p is one method of representing the damage inflicted on sheet elastomeric materials under specific aging times and temperatures. Q_p is the first explicit attempt made for linking the mechanical behavior of a sheet gasket material and its tightness behavior at elevated temperature.

A mechanical quality parameter Q_p greater than or equal to 0.5 is suggested as a primary guide for the acceptance of asbestos replacement gasket materials in order to evaluate their long-term service temperature. This is based on the fact that Q_p for asbestos materials remains over 0.5 for large values of exposures to time and temperature. Therefore, the candidate material is judged to be as good as asbestos reinforced sheets when its values for Q_p are over 0.5. For exposures where Q_p is less than 0.5, the candidate material fails.

Other possible quality parameters. Depending on which properties are used to express the gasket damage due to aging, several quantification tools can be developed and other criteria of acceptance can be established that should prevent the gasketed joint from excessive leaks during its service life under various conditions.

Considering the MTI tools developed essentially for the elastomeric sheet gaskets (mechanical quality parameter Q_p based on gasket load retention and residual tensile strength) other quantification tools can be developed on the basis of other physical gasket properties such as weight loss of matter, density change, or more generally compaction change of gasket components during aging.

It is clear that the development of general quantification tools applicable to a broad range of gasket materials is a praiseworthy objective. However, owing to the great variety of gasket concepts and gasket products available on the market, the quantification of changes in their mechanical and tightness properties during their real service life will always be a difficult job, and no one measure of mechanical or tightness quality has been found completely suitable for all types of gaskets. Studies are under way to investigate and extend the MTI findings and to develop more reliable "tools" to predict gasket behaviors and improve chances of selecting a "better" gasket for a given application.

In the early 1990s, an attempt based on the weight loss of gasket material was introduced to express the thermal degradation of elastomeric sheet gasket materials. In combination with load retention resistance and residual tensile strength, referred to as functional variables, the weight loss approach led to a better understanding of these materials.

In 1992, this approach was extended to flexible graphite based sheet materials. The degradation process of flexible graphite is mainly due to oxidation reactions. The rate of oxidation over time is accelerated by temperature. Past 700°F, an extensive decrease of creep and relaxation resistance over exposure time with temperature is observed for these materials while their postaging tensile strength is mostly that of the stainless steel inserts, which is not affected by the aging process. The correlation of these results with the weight loss measurements anticipates that oxidation will continue until flexible graphite disappears completely from the metallic insert, even at temperatures as low as 650 to 700°F. However, an extensive loss of relaxation resistance, around 50 to 60 percent, is obtained while only less than 5 percent weight loss is measured, especially at low temperatures around 700°F. As a result, there is no simple relationship between creep and relaxation and weight loss data, except for the low weight loss range.

The weight loss approach is an easy way to extend this concept to the idea that the density, or better, the compaction (or the porosity) change would be key properties to better understand the leakage behavior change of the gasket components during aging. Research work, at the time of this writing, is under way to define a suitable mechanical quality parameter based on compaction change during aging at elevated temperature of metal-reinforced flexible graphite based gaskets. This approach could be extended to other types of gasket materials or gasket types and lead to the definition of a quality parameter broad enough to support the idea of having the same mechanical quality basis for all gaskets.

Tightness quality parameters. It is useful to have a numerical measure of tightness quality for comparison and ranking purposes. Several tightness quality parameters have been defined for sheet gaskets and referred to as Q_t, Q'_t, and Q''_t. They are based on elevated-temperature leakage tests, referred to as HOTT/AHOT tests, described above and developed for MTI and PVRC. They are normalized to the performance of asbestos-reinforced gaskets. The first parameter is Q_t.

Specifically, Q_t is calculated as

$$Q_t = \frac{\log(T_{p\,\text{min}})/1.5 + f/1.5}{2}$$

- $T_{p\,\text{min}}$ is a key factor because it represents the residual gasket tightness performance after aging in the low-stress region of the HOTT test (T_p is the tightness parameter). For bolted flanged connections, it is in this low-stress region that catastrophic failures can occur.

- The slope f is that of the unload-reload cycles of the last part of the HOTT test. While the value of f is important because it is related to the tightness resilience of the gasket after aging, it is, however, considered to be of less importance than $T_{p\,min}$. In cases of high leak rates, the need to minimize the influence of the slope of a poorly performing gasket could be misleading. Thus f is normally limited to a maximum value of 3.

- The normalized factors of 1.5 are, respectively, the average test values of log $(T_{p\,min})$ and f obtained for traditional asbestos sheet materials.

Two other indexes have been found useful for tightness quality judgments. They are:

$$Q'_t = \frac{T_{p\,min}/32 + f/1.5}{2}$$

$$Q''_t = \frac{T_{p\,min}}{32}$$

In these equations, the 32 factor is $10^{1.5}$. This reflects the direct use of $T_{p\,min}$ instead of $\log(T_{p\,min})$. $T_{p\,min} = 32$ represents the value obtained for well-aged traditional asbestos sheet materials.

In general, Q_t, Q'_t, and Q''_t give consistent quality trends but Q''_t best exhibits the poor performance of a gasket candidate.

A pleasant convenience would be to have all gaskets on the same tightness quality basis, that is, to have available for other nonsheet gasket types including spiral-wound gaskets, a tightness quality index that is the same as that of the sheet gasket just described. Fortunately, currently available data support the idea of having the same tightness quality basis for all gaskets.

Tightness acceptance is currently judged on the basis of Q_t, Q'_t and Q''_t, each being equal to or greater than 1. In other words, the performance equals or exceeds that of the asbestos-based counterpart gasket.

Equivalent aged exposure parameters

From numerous mechanical ATRS/HATR tests performed on a wide variety of gasket products, the data obtained demonstrate a clear time-dependent effect on physical properties as they are quantified through the use of quality parameters. Thus there is a relationship between the amount of damage inflicted on a gasket material and the cumulative effects of exposure to temperature and time. These findings illustrate the possibility of combining time and temperature in a single relationship called an equivalent aged exposure parameter.

While it is possible to support the idea of having a general quality basis for all gasket products it is more difficult to imagine that there exists a general and single relationship between time and temperature. This is because gasket materials or gasket styles are so different in their manufacturing concept that a simple and general mathematical equation could not be broad enough to describe the effect of these two variables on the damage sustained by all types of gaskets. Many hundreds of ARTS/HATR or ARLA screen tests have now been completed on a variety of sheet materials ranging from traditional compressed asbestos, aramid, glass, and other fiber-reinforced elastomer substitutes to several varieties of PTFE and laminated flexible graphite metal-reinforced sheets as well as composite gaskets such as spiral-wound gasket styles or gaskets with metallic corrugated carriers. From this large data bank, several aging parameters have been introduced or are under development.

Equivalent aged exposure parameter A_e.[15] There are mathematical relationships which combine the properties that are considered critical as they can affect the good long-term performance of a gasketed joint in such a way that a good correlation is obtained for test data corresponding to different exposure time and test temperatures.

From the ATRS data available, multiple regression analysis between time, temperature, and these two mechanical properties represented by Q_r, Q_{tx}, or their product Q_p led to the development of a dimensionless equivalent exposure parameter A_e. A_e characterizes the combined effect of time and temperature on the damage affecting bound materials subjected to an oxidizing environment:

$$A_e = K(T-300) \times H^{0.2}$$

where T = exposure temperature, °F
 300 = reference temperature, °F, below which material sees no damage
 H = exposure duration, h
 K = constant determined on the arbitrary basis that $A_e = 100$ for a temperature of 800°F and an aging time of 1000 h
 = 0.0502 or 1/19.9054 when temperatures are expressed in °F
 0.2 = a standardized constant exponent for elastomer bound materials determined by best fit of experimental ATRS test results

The exposure parameter A_e provides a common basis for comparing the damage, expressed by Q_p, of one set of test conditions (time and temperature) to that inflicted by another set of time and temperature conditions. A_e is thus very useful for setting the test conditions for follow-up tests. It also provides a first rational basis to predict long-term performance of a sheet gasket material from short-term ATRS or HATR test data.

The choice of 800°F and 1000 h that makes $A_e = 100$ (for the MTI report[1]) was mostly arbitrary but closely linked to the mechanical performance of the nonasbestos materials that are tested during the MTI project. The choice of 100 is roughly twice the value where the quality Q_p for nonasbestos fiber materials tested in the MTI project goes to zero. There is no intent to say that asbestos reinforced sheet materials fail at least at $A_e = 100$. The possibility of $A_e > 100$ until failure is reached for these materials has been experimentally verified. In fact, experience shows that asbestos reinforced sheet products are widely and safely used at 750°F.

Any arbitrary change in the time-temperature combination that would make $A_e = 100$ (for example, 10 years and 1000°F so as to catch practically all service possibilities) will not result in different long-term estimates for the reinforced nonasbestos fiber elastomer materials.

Another parameter A_p for Elastomeric Sheet Gasket Materials.[25] In order to propose more accurate qualifying tools for any elastomeric type of gasket and extend their application to new gasket products coming on the market, some improvements have been implemented. One of the most important was to evaluate the deterioration due to temperature and time on the basis of weight loss of material due to thermal degradation.[6,7] This approach has led to the definition of an improved exposure parameter called A_p, which could be a more accurate predictive tool than A_e for elastomer bound materials.

A_p was derived from empirical equations representing gasket weight loss as a function of time and temperature for elastomeric sheet materials. Accounting for the rapid increase of the degradation rate that occurs above 330°C, A_p has been represented by two equations as follows:

$$A_p = 1000\, t^{0.6} \left(\frac{1}{T_{0'}} - \frac{1}{T_{0''}} \right) \quad \text{for } 360°F < T < 625°F$$

$$A_p = 1700\, t^{0.6} \left(\frac{1}{T_{0''}} - \frac{1}{T} \right) \quad \text{for } 360°F < T < 625°F$$

where $T_{0'}, T_{0''}$ = constants, °K
T = exposure temperature, °K

Aging parameter for metal-reinforced flexible graphite sheet gaskets.[34] Using the same approach as for the MTI, the following general relationship that is broad enough to cover both relaxation, creep, and weight loss data is now being considered:

$$A_f = K \times (T - T_0)^a \times H^b$$

where K = constant to be determined
T = test temperature
T_0 = threshold temperature below which there is no degradation
H = exposure time
a, b = constants to be determined by best fit of HATR relaxation and creep resistance with weight loss test data. From test data available presently, a value of 3 for exponent a and 0.4 for exponent b have been found to best represent the effect of time and temperature on the relaxation resistance and the weight loss of metal-reinforced flexible graphite sheet materials

The parameter A_f is a preliminary developmental tool that should lead to a final predictive tool to qualify and compare the hot performance of flexible graphite sheet gasket products and to determine their long-term operating temperatures. To be complete, leakage performance should be considered. A more comprehensive qualification protocol for flexible graphite sheet materials still needs more study on an acceptable load relaxation limit and on a maximum weight loss that can be tolerated to maintain an acceptable leakage behavior. As of the date of writing, tests required to achieve this goal are being conducted at Ecole Polytechnique in Montreal.

Qualification Guides or Protocols

Depending on the class of material, elastomer bound, flexible graphite, PTFE, or composite gaskets, the test scheme of the qualification guide will recommend a different mix of FIRS, ATRS, ARLA, and HOTT/AHOT test. Through the use of an aging parameter like A_e, A_p, or A_f, test conditions (temperature and time) are selected in order to qualify a material for a specific service temperature or to establish a recommended temperature limit. To judge the relative performance of a gasket, quality parameters like Q_p or weight loss and acceptance criteria have been established from comparative tests with asbestos sheet products.

Note that these predictions are based on the use of oxidizing media such as air. The various test programs and plant experience confirmed the validity of the mechanical ATRS test used as a screening tool and supplemented the overall adequacy of the gasket material qualification guide for asbestos substitute materials. For PTFE and flexible graphite based materials and for composite gaskets, the proposed qualification guides still need improvements to account for the specific behavior of these materials.

A draft specification scheme for elastomer bound sheet gaskets should be presented shortly to the ASTM Specification Committee for Gaskets, which is populated equally by users and manufacturers. Since the aging parameter A_e was derived specifically from tests on elastomer bound sheet materials, specification schemes for other types of gaskets need some adjustments or improvements before they can be submitted to official bodies. Research work is continuing on these subjects (see below).

Continuing Research Effort

The present North American research effort is aimed at developing the capability to predict and therefore improve the behavior of bolted flanged joints. In these environmentally sensitive times there is an urgent need for an approach in bolted joint design that considers leakage and makes the tightness of the joint a design criterion. The introduction in the ASME code of a bolt load calculation procedure based on the PVRC gasket constants will be a first step to solve this problem. However, for joints operating at elevated temperature, new design guidelines that are based on standardized qualification and test methods for gaskets are still to be finalized before further code modifications can be proposed. The following research programs are either currently under way or are to be undertaken very soon; they will help to achieve these goals:

- Long-duration screen tests at moderate temperatures on elastomer bound sheet materials (up to 12 months): to verify and establish the precision of long-term predictions based on short-term screen tests and the aging parameters A_e and A_p.
- Development of an aging parameter adapted to flexible graphite sheet materials: to improve the qualification scheme for flexible graphite products.
- Development of a Hot Blowout Test (HOBT) for gauging PTFE gasket tightness performance under extreme relaxation conditions: to improve the qualification scheme for PTFE based materials.
- Development of steam tests on full-scale gaskets and with a modified ARLA mechanical screening fixture. Steam and liquids are believed less harsh unless attack of chemical fluids is involved. Although a steam environment may be less harsh, a simple but representative screen test procedure is still needed to predict long-term service life of gaskets in

such an environment. The information obtained through these screening tests would supplement available test data resulting from tests performed by some gasket manufacturers or laboratories. The results should help determine what adjustments might be appropriate in the prediction of service temperatures for gaskets exposed to steam as compared to the gases we have used so far. We are presently working on another simple device, an upgraded version of the ARLA fixture, in which the gasket specimen would be aged not only under compressive stress, as in ARLA, but also under any type of pressurized fluids throughout the entire exposure to temperature and time. This new screening fixture could very well turn out to become the "sought after" standard.

- Development of an Emission Hot Tightness test (EHOT) to obtain pre- and postexposure gasket constants G_b, a, and G_g: to initiate gasket fugitive emission test program and establish leakage of organic fluids in flanged joints operating at moderate temperature (200 to 400°F).

References

1. ASME Boiler and Pressure Vessel Code, sec. VIII, div. 1, American Society of Mechanical Engineers, New York, 1989.
2. Payne, J. R.: "PVRC Flanged Joint User Experience Survey," *WRC Bulletin* 306, July 1985.
3. Bazergui, A.: "Short Term Creep and Relaxation Behaviour of Gaskets," *WRC Bulletin* 294, May 1984.
4. Jones, W. F., and B. B. Seth: "Evaluation of Asbestos Free Gasket Materials," ASME/IEE Power Generation Conference, Boston, October 1990, 90-JPGC/PWR-58.
5. Winter, J. R.: "Gasket Selection—A Flowchart Approach," for presentation at the 2d International Symposium on Fluid Sealing of Static Gasketed Joints, La Baule, France, September 18–20, 1990.
6. Design Division Problem No. XIII, "Re-evaluation of Gasket Factors Used in Flange Design," Long Range Plan for Pressure Vessel Research, 7th ed., *WRC Bulletin* 298, September 1984.
7. Raut, H. D., and G. F. Leon: "Report of Gasket Factor Tests," *WRC Bulletin* 233, December 1977.
8. Raut, H. D., A. Bazergui, and L. Marchand: "Gasket Leakage Behaviour Trends," *WRC Bulletin* 271, October 1981.
9. Bazergui, A., L. Marchand, and H. D. Raut: "Further Gasket Leakage Behaviour Trends," *WRC Bulletin* 325, July 1987.
10. Payne, J. R., and A. Bazergui: "More Progress in Gasket Testing—the PVRC Program," *1981 Proceedings Refining Dept.*, vol. 60, pp. 271–290, API, 1981.

11. Leon, G. F., and J. R. Payne: "An Overview of the US PVRC Research Program on Bolted Flanged Connections," *Proceedings of the ICPVT-6,* "Pressure Vessel Technology," edited by C. and R. Liu.
12. Bickford, J. H., K. H. Hsu, and J. R. Winter: "A Progress Report on US PVRC Joint Task Group on Elevated Temperature Behaviour of Bolted Flanges," *Proceedings of the ICPVT-6,* "Pressure Vessel Technology," edited by C. Liu and R. W. Nichols, vol. 1, *Design and Analysis,* pp. 249–266, Pergamon Press, New York, September 1988.
13. Hsu, K. H., J. R. Payne, J. B. Bickford, and G. F. Leon: "The US PVRC Elevated Temperature Bolted Flange Research Program," for presentation at the 2d International Symposium on Fluid Sealing of Static Gasketed Joints, La Baule, France, September 18–20, 1990.
14. Chao, R. C.: "Behaviour of Bolted Flanges at Elevated Temperature—Program Overview," *Proceedings, 1985 Pressure Vessel and Piping Conf.,* ASME PVP, vol. 98.
15. Payne, J. R., and A. Bazergui: "Evaluation of Test Methods for Asbestos Replacement Gasket Materials," MTI Publication 36, Materials Technology Institute of the Chemical Process Industries, Inc., 1990.
16. Payne, J. R., R. T. Mueller, and A. Bazergui: "A Gasket Qualification Test Scheme for Petrochemical Plants," parts I and II, presented at the 1989 PVP Conference, Honolulu, Hawaii, July 23–27, 1989, ASME PVP, vol. 158, pp. 53–79.
17. Rossheim, D. B., and A. R. C. Markl: "Gasket Loading Constants," *Mechanical Engineering,* vol. 65, p. 647, 1943.
18. Bazergui, A., J. R. Payne, and L. Marchand: "Effect of Fluid on Sealing Behaviour of Gaskets," *Proceedings, 10th International Conference on Fluid Sealing,* BHRA, Innsbruck, Austria, April 1984.
19. Bazergui, A., and L. Marchand: "PVRC Milestone Gasket Tests—First Results", *Welding Research Council Bulletin,* New York, WRC 292, February 1984.
20. Bazergui, A., L. Marchand, and H. D. Raut: "Development of Production Test Procedure for Gaskets," *Welding Research Council Bulletin,* New York, WRC 309, November 1985.
21. Bazergui, A., and G. Louis: "Predicting Leakage for Various Gases in Gasketed Joints," Society for Experimental Mechanics, 1987 Spring Conference on Experimental Mechanics, SEM, Houston, Tex., June 1987.
22. Payne, J. R.: "Draft 8—Standard Test Method for Gasket Constant for Bolted Joint Design," presented to ASTM Committee F3, February 1992.
23. Payne, J. R., A. Bazergui, and G. Leon: "New Gasket Factors—A Proposed Procedure," *Proceedings, 1985 Pressure Vessels and Piping Conference,* ASME/PVP, PVP, vol. 98.2, 1985.
24. Payne, J. R., G. Leon, and A. Bazergui: "Getting New Gasket Design Constant from Gasket Tightness Data," Special Supplement, Experimental Techniques, Society of Experimental Mechanics, November 1988.
25. Marchand, L., M. Derenne, and A. Bazergui: "Weight Loss Correlation for Sheet Gasket Materials," *ASME Journal of Pressure Vessel Technology,* in press.
26. Marchand, L., A. Bazergui, and M. Derenne: "Recent Developments in Elevated Temperature Gasket Evaluation," 13th International Conference on Fluid Sealing, Brugge, Belgium, Apr. 7–9, 1992.

27. Bazergui, A., and J. R. Payne: "On Elevated Temperature Behaviour of Gaskets," *Proceedings ICPVT-6, Pressure Vessel Technology,* vol. I, *Design and Analysis,* Pergamon Press, New York, September 1988.
28. Payne, J. R., M. Derenne, and A. Bazergui: "A Device for Screening Gasket Materials at Elevated Temperature," *Proceedings, 11th Fluid Sealing Conf.,* BHRA, Elsevier Applied Science Publisher, Cannes, France, April 1987.
29. Marchand, L., A. Bazergui, and M. Derenne: "The Influence of Thermal Degradation on Sealing Performance of Compressed Sheet Gaskets with Elastomer Binder, Part 1: Experimental Methods," 2d International Symposium on Fluid Sealing of Static Gasketed Joints, CETIM, La Baule, France, September 1990.
30. Derenne, M., J. R. Payne, L. Marchand, and A. Bazergui: "On the Fire Resistance of Gasketed Joints," *Welding Research Council Bulletin,* New York, in press.
31. Bazergui, A., L. Marchand, and J. R. Payne: "Development of a Hot Tightness Test for Gaskets," *Proceedings, 11th Fluid Sealing Conf.,* BHRA, Elsevier Applied Science Publisher, Cannes, France, April 1987.
32. Bazergui, A., L. Marchand, and J. R. Payne: "Development of Tightness Test Procedures for Gaskets in Elevated Temperature Service," *Welding Research Council Bulletin,* WRC 339, December 1988.
33. Gasket Testing Laboratory of the GACM: "Testing Gasket Performance at Room and Elevated Temperature," 1992.
34. Derenne, M., and L. Marchand: "Elevated Temperature Tests on Metal-Reinforced Flexible Graphite Materials," *PVRC Progress Report,* May 1993.

Chapter 5

The Gasket and the Application

Gasket and Joint Diagram

The effects of unit operation on a gasket's properties can best be noted with a review of the gasket and joint diagram. This is explained as follows: Figure 5.1 depicts an idealized linear relationship of a gasket's load versus deflection of a gasket under compression.

Figure 5.2 depicts an elastic curve of load versus deflection for a gasketed joint. In actuality this represents the combination of bolt stretch and compression of the mating flanges.

Figure 5.3 combines these two elastic curves.

Upon operation of the unit, a hydrostatic end force is exerted on the gasketed joint. The effect of this force on the gasket and joint diagram is shown in Fig. 5.4. The length of line CD represents the hydrostatic end force. Line BC represents the increase in bolt stretch as a result of this force. Line BD shows the corresponding reduction in gasket loading.

As noted earlier, upon operation of a unit, the gasket undergoes a stiffening effect. This is represented by line EB in Fig. 5.5.

Line FG, which equals line CD in length, shows the new loads on the bolts and gasket. Note that the increase in bolt load is less while the load on the gasket is reduced more in the case of

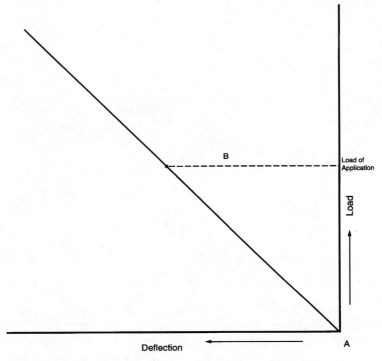

Figure 5.1 Load deflection for a gasket.

the final gasket. In addition to the stiffening effect, the gasket undergoes relaxation as the unit operates. The change in the gasket and joint diagram due to this relaxation is shown in Fig. 5.6. The static load of the application has been reduced from B to I as a result of the relaxation. The dynamic load has been reduced from G to K for the same reason.

Figure 5.7 is a compilation of all these previous gasket and joint diagrams and shows all the effects in one figure. The following is a compilation of the nomenclature associated with this figure.

$$OB = \text{bolt stress versus elongation}$$

$$AB = \text{initial gasket stress compression}$$

$$CD = \text{hydrostatic end force}$$

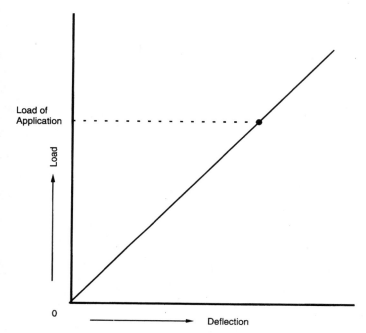

Figure 5.2 Load deflection for a bolt.

C = increase in bolt load due to CD

D = reduced stress on gasket due to CD

EB = final gasket stress compression

$FG = CD$

F = increase in bolt load due to FG

G = reduced stress in gasket due to FG

Note that the sealing stress on the gasket has been reduced from the initial level at point D to point G as a result of a change in the gasket spring rate during operation. The effect of relaxation of the gasket material can also be noted in this figure.

HI = same spring rate as EB

BI = relaxation of the gasket

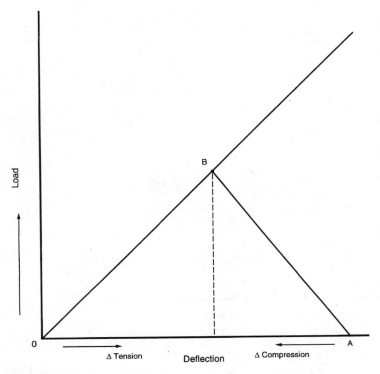

Figure 5.3 Gasket and joint diagram.

$$JK = FG = CD$$

J = increase in bolt load due to JK

K = reduced stress on gasket due to JK

The length and spring rate of the flange bolts also have an effect on seal maintenance. As illustrated in Fig. 5.8, for a given amount of compression set, the load loss of a high spring bolt is greater than that for a bolt having a lower spring rate. Therefore, to reduce the bolt spring rate for least load loss, the diameter should be as low as possible and the bolt should be as long as possible.

Gasketing may become extremely complicated when a variety of media are to be sealed simultaneously; for example, the head gasketing of an internal combustion engine requires the

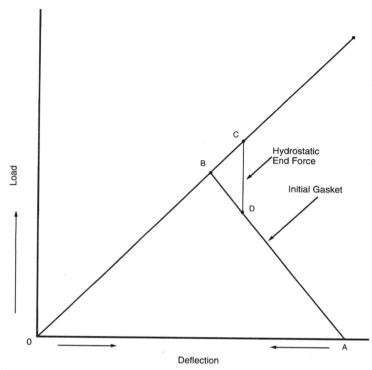

Figure 5.4 Identification of the hydrostatic end force.

sealing of high-pressure combustion gases as well as crankcase oil and water containing antifreeze. The material must be impervious to gases, fluids, and temperatures involved and, in order to avoid the need for retorquing, must take only a minimal compression set. The coolant and oil passageways require lower but sufficient loading to achieve complete sealing. Head bolt size, spacing, and length, head and block stiffness, combustion pressure, and engine speed also enter as sealing variables.

Often a value analysis approach to gasketing may result in the lowest overall jointing cost. One must recognize that the gasket is only one part of the jointed system, and often, by using a more sophisticated gasket, significant savings may be achieved on the castings, stampings, or related hardware of the gasketed joint.

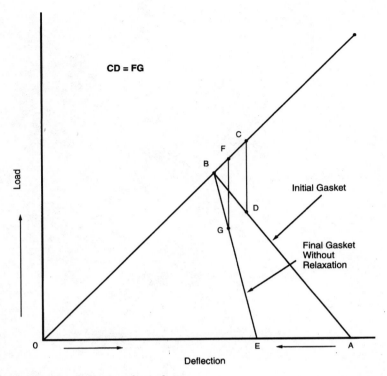

Figure 5.5 Stiffening of a gasket.

Application Information

In order for a gasket engineer to design the best gasket for a specific application, various details of the application must be known. Data on the fastening system flanges and sealing media are required for proper analysis and design. Figure 5.9 is a sample data sheet that can be filled out and sent to the gasket engineer.

Gasket Installation

An installation is only as good as its gasket; likewise, a gasket is only as good as its installation. The following are some recommendations associated with gasket installation:

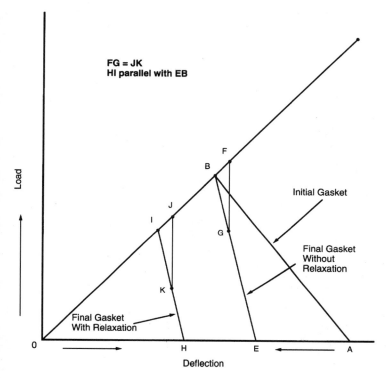

Figure 5.6 The effect of gasket relaxation.

- Be sure that mating surfaces are clean and in specification with regard to finish.
- Clean bolt holes in mating flanges.
- Check gasket for damage before installing it.
- Make certain the gasket fits the application.
- Specify lubricated bolts. Bolt threads and the underside of the bolt head should be lubricated.
- Make certain that the bolts do not bottom out in the mating flange.
- Specify the torque level and use of a torque wrench.
- Specify the torquing sequence. In addition to the sequence, two or three stages of torque before reaching specified level are recommended.

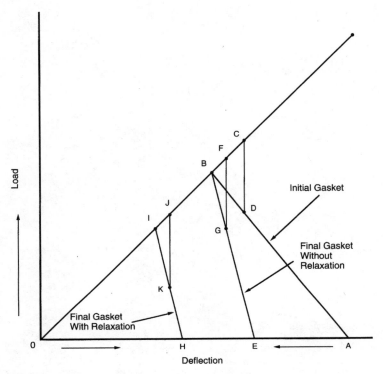

Figure 5.7 Complete gasket and joint diagram.

Sealing Enhancements

In some cases, a gasketed joint's sealing requirements are such that produced gasket sheet materials cannot accommodate them. In these cases, the gasket fabricator utilizes any number of sealing aids or enhancements to improve the gasket to meet these requirements. Some of these sealing enhancements are:

Saturating

The voids in the gasket material can be filled with a number of saturating chemicals. The saturants, in addition to filling the voids, can add improved heat and chemical resistance to the gasket. They can also alter the physical properties of the material. Some saturants are cured during the fabrication of the gaskets while others are processed so that they cure during

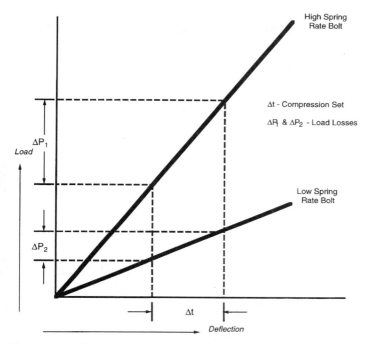

Figure 5.8 Effect of bolt spring rate on load loss.

operation of the gasketed joint. Common saturants include silicones and phenolics.

Coating

A wide variety of gasket coatings are available in the marketplace. The coatings are used for a number of different reasons. Some saturate into the material and fill voids while most others are used for surface sealing, antifret, or antistick purposes. Some coatings are solvent-based and others are water-based.

The following is a listing of some of the reasons to incorporate a coating on a gasket:

- Improve surface sealing.
- Improve antifret characteristics. Antifret is reduction of scrubbing of the gasket due to flange shearing motions.
- Reduce or eliminate sticking upon removal.

GASKET APPLICATION REQUIREMENTS

NAME: _____ TITLE: _____

CUSTOMER: _____ ADDRESS: _____

CITY: _____ STATE: _____ ZIP: _____ PHONE: _____

TYPE OF UNIT: _____ FAX: _____

GASKET SEALING APPLICATION: _____

PART NO. REF.: _____

APPLICATION DATA

BOLT DIA.: _____ BOLT TORQUE OR LOAD: _____ BOLT GRADE: _____

NUMBER OF BOLTS: _____ TOTAL LOAD ON APPLICATION: _____ BOLT LENGTH: _____

CURRENT GASKET MATERIAL AND/OR CONSTRUCTION: _____

GASKET THICKNESS REQUIREMENTS: _____

ENVIRONMENT DATA

TYPE OF MEDIUM TO SEAL: _____

PRESSURE RANGE: NORMAL OPERATING _____ HIGH _____ LOW _____

TEMPERATURE RANGE: NORMAL OPERATING _____ HIGH _____ LOW _____

FLANGE DATA

	FLANGE A	FLANGE B		FLANGE A	FLANGE B
MATERIAL:	_____	_____	THICKNESS:	_____	_____
FINISH:	_____	_____	FLATNESS:	_____	_____

DATA ENCLOSED FOR EVALUATION

☐ CARBON IMPRESSION ☐ GASKET PRINT ☐ TORQUE SPECS
☐ PRESENT GASKET ☐ FLANGE PRINTS ☐ GASKET SKETCH
☐ TEST PROCEDURES ☐ ASSEMBLY SPECIFICATIONS
☐ MISC.

Figure 5.9 Gasket application data sheet.

- Provide sticking or tack for ease of assembly.
- Provide gasket with a barrier coat for subsequent printing (printing is discussed later).
- Reduce sticking of gasket to gasket during processing and shipping.
- Provide color to the gasket for identification reasons.

The following are some coatings used for various reasons:

Molybdenum disulfide. This is a high-temperature lubricating release coating with excellent wear life and resistance to solvents and fluid. It can be used for continuous duty, sliding friction applications.

Teflon.* This coating is soft and malleable, thereby improving surface sealability. It is nonadhering, resulting in clean removal. It provides for movement between flanges.

Molybdenum disulfide and Teflon. This coating is applied primarily to gaskets to prevent fretting. It is a malleable coating that burnishes into the flange surface. This aids sealability and the ability to accommodate motion. It is nonadhering, allowing easy and clean removal.

Epoxy and phenolic. This coating has excellent release properties. It also can be used as a barrier coat, as noted earlier.

Silicone A. This silicone coating is a soft elastomeric, conformable sealing coating applied for improved surface sealing. In addition, it has excellent release properties at disassembly.

Silicone B. This silicone coating is a saturant and is applied to nonmetallic gaskets to improve fluid sealing through the material.

Silicone and mica. This silicone coating is the same as silicone A. The addition of mica improves both the surface sealing and release properties.

Teflon and mica. This coating is primarily an antistick but also is good for surface sealing. It is soft and malleable, thereby aid-

---

*Teflon is a registered trademark of the DuPont Co.

ing sealability. It is nonadhering, resulting in clean removal. Its antifret properties provide for movement between the flanges.

Aluminum epoxy ester. This coating is applied to provide a highly conformable sealing surface. It is resistant to water, coolants, and oil.

Neoprene or nitrile. These coatings are elastomeric and provide excellent surface sealing. Because they will adhere to the mating flanges, an antistick coating should be used if antistick properties are desired.

Copper or aluminum. These coatings are soft and conformable and are normally used with gaskets to improve surface sealing.

Mica. This is an antistick coating.

Nitrile. This coating is a pressure-sensitive adhesive. It is applied to gaskets along with a mylar film. The gasket installer removes the film and attaches the gasket to the flange applying pressure.

Polyacrylic. This coating is used for improved surface sealing, primarily for oil applications.

Table 5.1 depicts the generic characteristics of various surface coatings used for gaskets. Gasket manufacturers should be consulted to confirm that their coatings comply with these characteristics.

Eyeletting

Metal eyelets are used at port openings to (1) protect the gasket material from the sealed media, and/or (2) provide high sealing stress on the eyelet. In some cases, eyelets can also be used at bolt holes to reduce distortion of the gasketed joint (Fig. 5.10).

Common metals used for eyelets are copper, steel, and stainless steel. Eyelet dimensions must be restricted to ensure that fracture and/or wrinkles do not occur during their processing. Table 5.2 depicts the recommended dimensions to be maintained. The numbers are based on 0.010-in copper.

TABLE 5.1 Surface Coating Functions

Type	Improve Surface Seal	Anti-Fret Protection	Easy Release Upon Disassembly
Polytetrafluoroethylene (PTFE)	+	+	+
Silicone (VMQ)	+	+	+
Molybdenum Disulfide (MoSz)	+	+	+
Moly/PTFE Blends	+	+	+
Inorganics			
Graphite	+/0	+	+
Mica	0	0	+
Talc	0	0	+
Epoxy/Phenolics	+/0 *	0	+
Enamels	+ *	0	+
Nitrile (NBR)	+ *	–	0/– **
Neoprene (CR)	+ *	–	0/– **

KEY: + = Excellent
0 = Fair
– = Unsuitable

* Runs, sags, and uneven application may cause surface leakage

** Can be improved by application of a good releasing "surface coating"

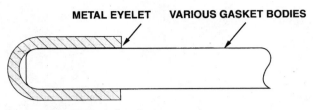

Figure 5.10 Eyeletted gasket.

Example Assume a gasket is to be 1/16 in thick and a minimum overlap of 3/32 in is desired. The table shows that the radius for the sharpest turn should be not less than 7/16 in.

Use of different thicknesses and/or materials will require different dimensions. For these cases, the overlaps should be reduced or the radius should be increased by the following ratios:

Copper thickness, in	Ratio	Steel thickness, in	Ratio
0.006	0.9	0.006	0.75
0.010	1.0	0.010	0.85
0.015	1.15	0.015	0.95
0.020	1.20		

Example 1 0.010 in copper can be formed to 1/8-in overlap on a 1-in radius. 0.006-in steel can be formed 0.125 (1/8 in)×0.75 = 0.094-in overlap.

Example 2 0.010-in copper requires a 1-in radius for 1/8-in overlap. 0.006-in steel requires 1 in×100/75 = 1.33-in radius.

Metal reinforcement

Metal is also used to reinforce nonmetallic gasket materials. The material's facing may be the compressed variety or the beater additive type.

Traditionally, perforated or upset metal cores have been used to support gasket materials. A wide variety of different designs have been utilized for production. Rectangular perforations are most common. Size of the perforations and their frequency in a given area are the usual specified parameters.

Recently, adhesives have been developed that also permit the use of an unbroken metal core to reinforce gasket facings. Laminated composites of this type have certain characteristics that are desired in particular gaskets. These characteristics are discussed later. The demanding environment of various gasket-

TABLE 5.2 Recommended Eyelet Dimensions

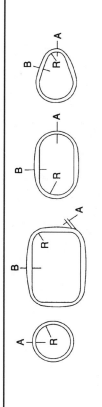

T = Thickness of Gasket   A & B = Overlap Widths

R \ T		1/8		3/16		1/4		5/16		3/8		7/16		1/2		9/16		5/8		11/16		3/4		13/16		7/8		15/16		1	
		A	B	A	B	A	B	A	B	A	B	A	B	A	B	A	B	A	B	A	B	A	B	A	B	A	B	A	B	A	B
1/32		1/16	1/8	1/16	1/8	3/32	1/8	3/32	1/8	3/32	5/32	3/32	5/32	3/32	5/32	1/8	5/32	1/8	5/32	1/8	5/32	1/8	5/32	1/8	5/32	1/8	5/32	1/8	5/32	5/32	5/32
3/64		1/16	1/8	1/16	1/8	3/32	1/8	3/32	1/8	3/32	5/32	3/32	5/32	3/32	5/32	1/8	5/32	1/8	5/32	1/8	5/32	1/8	5/32	1/8	5/32	1/8	5/32	1/8	5/32	5/32	5/32
1/16		1/16	1/8	1/16	1/8	3/32	1/8	3/32	1/8	3/32	5/32	3/32	5/32	3/32	5/32	1/8	5/32	1/8	5/32	1/8	5/32	1/8	5/32	1/8	5/32	1/8	5/32	1/8	5/32	5/32	5/32
5/64		1/16	1/32	1/16	1/8	1/16	1/8	1/16	1/8	1/8	1/8	3/32	5/32	3/32	5/32	1/8	5/32	1/8	5/32	1/8	5/32	1/8	5/32	1/8	5/32	1/8	5/32	5/32	5/32		
3/32						1/16	1/8	1/16	1/8	1/8	1/8	1/8	1/8	1/8	1/8	1/8	1/8	3/32	1/8	3/32	1/8	3/32	1/8	3/32	1/8	3/32	1/8	3/32	1/8	1/8	5/32
1/8								1/8	1/8	1/8	1/8	1/8	1/8	1/8	1/8	3/32	1/8	3/32	1/8	3/32	1/8	3/32	1/8	3/32	1/8	3/32	1/8	3/32	1/8	1/8	5/32

ed joints makes a high-strength, metal adhesive-facing bond necessary for the laminated composite. Figures 5.11 and 5.12 depict perforated and unbroken metal core composites, respectively.

Metal-reinforced gaskets are classified as laminated composite gasket materials (LCGM) by ASTM. The classification system consists of a line call-out that describes the use, composition, and combining method. Various sections of Standard ASTM F868, which covers composite gaskets, follow:

ASTM F868

Standard Classification for Laminated Composite Gasket Materials. **Scope:** This classification system provides a means for specifying or describing pertinent properties of commercial laminated composite gasket materials (LCGM). These structures are composed of two or more chemically different layers of material. These materials may be organic or inorganic, or combinations with various binders or impregnants. Gasket coatings are not covered since details thereof are intended to be given on engineering drawings or as separate specifications. Commercial materials designated as envelope gaskets are excluded from this standard.

Since all the properties that contribute to gasket performance are not included, use of this classification system as a basis for selecting LCGM is limited.

This standard may involve hazardous materials, operations, and equipment. This standard does not purport to address all

Figure 5.11 Perforated core composite.

Figure 5.12 Unbroken metal core composite.

the safety problems associated with its use. It is the responsibility of the user of this standard to establish appropriate safety and health practices and determine the applicability of regulatory limitations prior to use.

Terminology: *Descriptions of Terms Specific to This Standard:*

board: The term board is used in the context of a thick (generally greater than 0.060 in) and rigid nonmetallic material, often purchased in sheet or strip form.

composite gasket material: A gasket structure composed of two or more different materials joined together in flat, parallel layers.

Significance and Use: This classification is intended to encourage uniformity in reporting properties; to provide a common language for communications between producers and users; to guide engineers and designers in the use, construction, and properties of commercially available materials; and to be versatile enough to cover new materials and test methods as they are introduced.

Basis of Classification: This classification is based on the principle that LCGM should be described, insofar as possible, in terms of use, composition, combining method, and specific physical and mechanical characteristics. Thus, users of gasket materials can, by selecting different combinations of materials and properties, define various parts. Suppliers, likewise, can report uses, composition, and properties of available products.

Numbering System: To permit line call-out of the description mentioned, this classification establishes letter or number symbols to describe use, composition, and physical properties and performance levels of certain properties.

In specifying or describing gasket materials, each line callout should include the number of this system and a number and letter series describing the use, composition, and combining method plus suffix call-out, as shown in Table 5.3.

To further specify or describe gasket materials, each line callout may include one or more suffix letter-numeral symbols, as listed in Table 5.4.

Physical and Mechanical Properties: Gasket materials identified by this classification should have a number and

TABLE 5.3 Basis of Classification

First digit Typical end-use	Letter group Composition (Component material)	Second digit Combining method
0. Not specified	N. Not specified	0. Not specified
1. Carburetor, engine	B. Board	1. Tanged perforation
2. Intake manifold, engine	M. Metal	2. Chemical bond
3. Exhaust manifold, engine	F. Classification F 104 material	3. Tanged perforation plus chemical bond
4. Cylinder head, engine	R. Rubber Classification D 2000	4. Grommets
5. Transmission, engine	P. Plastics	5. Overlap
6. Ducts and piping	T. Textiles	6. Bonded and vulcanized
7. Compressors	S. As specified	9. As specified
9. As specified		

Suffix designation
Any specific test requirement
Letters represent types of tests
Numbers represent values

NOTE--This classification is intended to be open-ended with a two-digit plus letter group call-out. The letters in the group for a given composite gasket material will be those representing the layers in order.
Example: 4 FMF1; F=F112440; M-Specification A 109

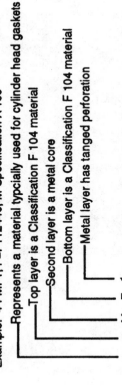

Represents a material typcially used for cylinder head gaskets
 Top layer is a Classification F 104 material
  Second layer is a metal core
   Bottom layer is a Classification F 104 material
    Metal layer has tanged perforation

4 F M F 1

TABLE 5.4 Supplementary Physical and Mechanical Characteristics

Suffix Symbol	Supplementary Characteristics
A9	*Sealability* characteristics determined as agreed upon between supplier and user.
C9	*Compressibility* characteristics as agreed upon between producer and user.
D9	*Release* characteristics determined as agreed upon between supplier and user.
E00 through E99	*Weight and thickness change after immersion in ASTM Fuel B* shall be determined in accordance with Test Method F 146. *Weight Increase* shall not exceed the standard rating number indicated by the first numeral of the two-digit number of the E-/symbol. *Thickness Increase* shall not exceed the standard rating number indicated by the *second* numeral of the E-symbol.

Percent Weight Increase
(first numeral)

E0_ = not specified
E1_ = 10
E2_ = 15
E3_ = 20
E4_ = 30
E5_ = 40
E6_ = 50
E7_ = 60
E8_ = 100
E9_ = as specified

Percent Thickness Increase
(second numeral)

E_0 = not specified
E_1 = 0-5
E_2 = 0-10
E_3 = 0-15
E_4 = 5-20
E_5 = 10-25
E_6 = 15-35
E_7 = 25-45
E_8 = 30-60
E_9 = as specified

| G00 through G99 | *Weight and thickness change after immersion in ASTM #3 Oil* shall be determined in accordance with Test Method F 146. *Weight Increase* shall not exceed the standard rating number indicated by *first* numeral of the two-digit number of the G-/symbol. *Thickness Increase* shall not exceed the standard rating number indicated by the *second* numeral of the G-/symbol. |

G Table:

Percent Weight Increase
(first numeral)

G0_ = not specified
G1_ = 10 %, max
G2_ = 15 %, max
G3_ = 20 %, max
G4_ = 30 %, max
G5_ = 40 %, max
G6_ = 60 %, max
G7_ = 80 %, max
G8_ = 100 %, max
G9_ = as specified

Percent Thickness Increase
(second numeral)

G_0 = not specified
G_1 = 0-15 %
G_2 = 5-20 %
G_3 = 10-25 %
G_4 = 15-30 %
G_5 = 20-40 %
G_6 = 30-50 %
G_7 = 40-60 %
G_8 = 50-70 %
G_9 = as specified

TABLE 5.4 Supplementary Physical and Mechanical Characteristics (*Cont.*)

H1 through H9 — *Creep relaxation* characteristics shall be determined in accordance with Test Method F 38. Loss of stress at the end of the test shall not exceed the amount indicated by the numeral of the H-/symbol.

H1 = 10%
H2 = 15%
H3 = 20%
H4 = 25%
H5 = 30%
H6 = 40%
H7 = 50%
H9 = as specified

K0 through K9 — *Thermal conductivity* characteristics shall be determined in accordance with Practice F 433. The K-factor obtained in W/m·K (Btu-in./h·ft^2·°F), shall fall within the ranges indicated by the numeral of K-/symbol.

K1 = 0. to 0.09 (0 to 0.65)
K2 = 0.07 to 0.17 (0.50 to 1.15)
K3 = 0.14 to 0.24 (1.00 to 1.65)
K4 = 0.22 to 0.31 (1.50 to 2.15)
K5 = 0.29 to 0.38 (2.00 to 2.65)
K6 = 0.36 to 0.45 (2.50 to 3.15)
K7 = 0.43 to 0.53 (3.00 to 3.65)
K8 = 0.50 to 0.60 (3.50 to 4.15)
K9 = as specified

L9 — *Laminated bond* characteristics as agreed upon between producer and user.
X9 — *Crush-extrusion* resistance characteristics as agreed upon between producer and user.
Y9 — *Coating(s)* per drawing details.
Z0 through Z9 — Any other properties per drawing details.

[a] On engineering drawings or other supplements to this classification.

letter call-out for end-use and construction indicated in Table 5.3 and additional properties by a letter-numeral call-out shown in Table 5.4.

Thickness Requirements: Gasket materials identified by this classification should conform to the thickness specified on the gasket drawing, or on the order.

The thickness of individual components of the composite may be specified on the drawing, where necessary, and where component can be measured.

Sampling: Specimens should be selected from finished gaskets or sheets of suitable size, whichever is the more practicable. If finished gaskets are used, the dimensions of the sample and any variations from the method must be reported.

Sufficient specimens should be selected to provide a minimum of three determinations for each test specified. The average of the determinations should be considered as the result.

Conditioning: Prior to all tests, specimens should be conditioned as follows:

- When all Classification F104 layers of the composite are of the same "type," condition per that type.

- When the layers of the composite are of different Classification F104 "types," the composite should be conditioned 22 h in a controlled humidity room, or in a closed chamber containing air at 70 to 86°F and 50 to 55 percent relative humidity.

- Other conditioning may be as agreed upon between producer and user.

Test Methods: The test methods are indicated in Table 5.3 under each suffix symbol when appropriate.

Table 5.5 depicts the suffix letters and test methods for laminated composite gasket materials.

To show the importance of a reinforcement for a gasket facing, a number of tests were conducted on plain and reinforced facings. The unnumbered table on p. 174 depicts the various constructions involved in this testing.

TABLE 5.5 Suffix Letters and Test Methods for Laminated Composite Gasket Materials

Suffix symbol	Test description	0	1	2
A	Sealability			
C	Compressibility			
D	Release			
E—first number	ASTM fuel B, ASTM F146 % weight increase	NS	10	15
E—second number	ASTM fuel B, ASTM 146 % thickness increase	NS	0–5	0–10
G—first number	ASTM no. 3 oil, ASTM F146 max T weight increase	NS	10	15
G—second number	ASTM no. 3 oil, ASTM F146 % thickness increase	NS	0–15	5–20
H	ASTM F38 % stress loss	NS	10	15
K	Thermal conductivity, W/M °K ASTM F433	NS	0–0.9	0.07–0.17
L	Laminated bond characteristics			
X	Crush-extrusion resistance			
Y	Coatings			
Z	Drawing detailed properties			

NS = not specified
AS = as specified and agreed upon between supplier and user

	Composite thickness, in		
Construction	0.032	0.048	0.062
Plain material	0.032	0.048	0.062
Perforated core	0.012	0.020	0.028
	0.008	0.008	0.008
	0.012	0.020	0.028
Laminated (unbroken metal core)	0.012	0.012	0.012
	0.008	0.024	0.040
	0.012	0.012	0.012

These constructions were tested to determine the properties of compression-deflection, stress-relaxation, crush and extrusion, and sealability.

Compressive deflection versus thickness. As can be noted in Fig. 5.13, the compression-deflection properties of the supported materials are considerably stiffer than those of the plain material. This is true even when the same amount of compressible material is inherent in each construction. The reason for

3	4	5	6	7	8	9
						AS
						AS
						AS
20	30	40	50	60	100	AS
0–15	5–20	10–25	15–35	25–45	30–60	AS
20	30	40	60	80	100	AS
10–25	15–30	20–40	30–50	40–60	50–70	AS
20	25	30	40	50		AS
0.14–0.24	0.22–0.31	0.29–0.38	0.36–0.45	0.43–0.53	0.50–0.60	AS
						AS
						AS
						AS

this is that the supported materials are restrained from lateral flow by the support metal. Owing to this restraint, it can also be expected that the stress-relaxation properties of supported materials will be better than those of unsupported facings.

Stress relaxation. The table below depicts the results of stress relaxation testing of the constructions.

Stress Relaxation* (Percent Loss versus Thickness)

Specimen	Thickness, mm			Test temperature
	0.8	1.2	1.6	
Plain	20.2	24.3	29.7	
Perforated	18.6	21.6	27.7	At 212°F
Laminated	18.5	21.3	20.0	
Plain	33.6	42.3	47.3	
Perforated	33.0	47.1	44.8	At 300°F
Laminated	35.2	34.5	37.5	

*ASTM F38 Method B, 22 h at noted temperature and 3000 psi pressure.

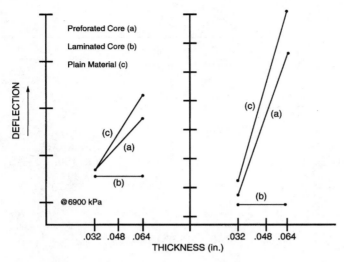

Figure 5.13 Compressive deflection versus thickness of various constructions or composites.

The stress-relaxation results can be explained as follows: Perforated core materials rely on increased thickness of facing material to achieve increased composite thickness. Production limitations set a maximum thickness of the perforated core at 0.010 in. Generally, the thickness is 0.008 in. Laminated constructions utilizing unbroken metal cores can maintain constant facing thickness while varying core thickness to obtain a desired overall thickness. Gasket material thickness is preferably kept at a minimum for torque retention purposes. It must be thick enough, however, to provide accommodation of the warpage conditions of the flanges being sealed.

Thus, at the 0.032-in thickness, composition of the perforated and laminated (unbroken metal core) constructions is the same. With thicker gasket combinations, the perforated core requires thicker materials while the laminate utilizes thicker metal cores. Both constructions have better stress-relaxation properties than the unsupported material. Torque retention will be improved when stress-relaxation properties are better. It has been documented that relaxation and torque loss are directly related to thickness of compressible material.

Crush and extrusion. Similar results were obtained in the crush-extrusion testing of these constructions. Both of the supported (perforated and laminated) materials resisted crushing and extruding better than the plain material. Table 5.6 depicts the results of this testing. The specimens were prebaked at 375°F, then aged as noted and subjected to the specified pressure for 2 min at 300°F. Their area was measured and compared to the original area, which was set at a figure of 100. Better crush-extrusion properties reflect improved crush resistance in cases of highly concentrated loads. These loads occur under bolts in weak flanges and sometimes in gaskets where metal eyelets are utilized.

Sealability. Sealability of the plain (c) and laminated (b) construction is similar, as can be noted in Fig. 5.14. At low clamp load, both are better than the perforated construction (a). As clamping stress level is increased, the sealability of perforated construction approaches that of the others. This follows from

TABLE 5.6 Crush and Extrusion Test Results

		\(Area of Original Specimens = 100\)					
		THICKNESS (in.)					
		.032		.048		.064	
PRESSURE →		A	B	A	B	A	B
PLAIN	Aged in 10W40 Oil	117	138	126	153	143	185
	Aged in Glycol/Water	113	169	155	185	202	215
PERFORATED	Aged in 10W40 Oil	117	126	120	153	164	194
	Aged in Glycol/Water	128	143	149	189	179	213
LAMINATED	Aged in 10W40 Oil	113	130	109	143	115	130
	Aged in Glycol/Water	123	138	111	149	123	138

PRESSURE A = 30,000 psi
PRESSURE B = 50,000 psi

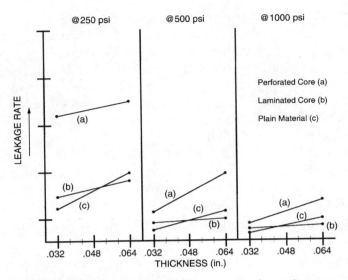

Figure 5.14 Sealability versus thickness for the various constructions.

the fact that the perforations are possible leakage paths. At low clamp load the facing may not be pressed tightly against the perforations and leakage can result.

Note that the plain material is slightly better than the laminate construction in the case of the 0.032-in construction. This is probably due to the slight burr that absorbs load on the blanked, laminated specimen. At 0.048-in and greater thicknesses, the laminate construction is slightly better than the plain laminate as there is less gasket material in the laminate for leakage to occur.

Radial strength. Increased radial strength in gaskets is manifest with unbroken steel core versus perforated core, even with core thicknesses being equal. Of course, the thicker, unbroken steel cores provide improved strength. This strength factor can be important when sealing high pressures. In addition, the unbroken steel core construction results in stronger thin-wall sections that occur on many gaskets. This also enhances the handling characteristics of these gaskets.

Thickness and torque loss. Another benefit of the unbroken steel core versus the perforated steel core construction is associated with the torque loss characteristics of multiple thickness gaskets. To maintain torque and minimize distortion, the gasket should contain the minimum amount of compressibility. Sufficient compressibility, however, must be inherent in the gasket for adequate seal. More than this amount will result in higher stress relaxation and higher subsequent torque loss. To keep compressibility as low as possible, the least amount of gasket material is desired. This construction is regardless of the gasket's thickness since the metal core thickness can be changed while the facing material thickness is kept constant.

In the case of the perforated core construction, changing the metal thickness affects the characteristics of the composite. That is, thicker metals change the relative percentage of metal and facing within the facing layer. As noted earlier, there is a limit on the thickness of the metal that can be perforated.

Embossing. Another advantage for the unbroken metal core construction is in the case of embossing (Fig. 5.15). The unbroken metal core emboss is more rigid than the emboss of the

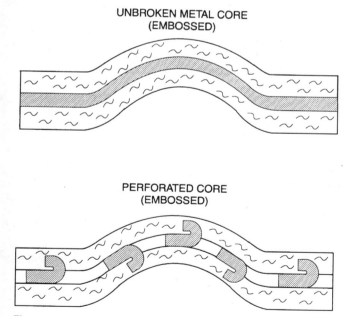

Figure 5.15 Unbroken and perforated metal core emboss.

perforated core. It supports a higher load and exhibits higher recovery characteristics. Figure 5.16 depicts the load-deflection and recovery characteristics of the two embossed constructions. These characteristics can be widely varied by using different material thicknesses and tempers as well as different metals.

Stress-sensitive film impressions were used to determine the stress distribution of both the perforated and unbroken metal core embossed gaskets. Methods of stress distribution testing are included later. The impressions were made for one and two applications of load. Improved stress levels were maintained by the unbroken metal core emboss, as evidenced by the similarity of the carbon impressions for both applications of load as compared to the perforated core impressions.

Printing

Another enhancement which is used to improve sealing of gasket materials is the depositing of an elastomeric bead on the gasket surface. Figure 5.17 depicts the elastomeric bead deposited on a gasket body.

The most common method is the silk screening process. In this process, elastomeric beads are strategically located in the critical sealing portions of the gasket. The beads may also be deposited by other means, but the most common is silk screening.

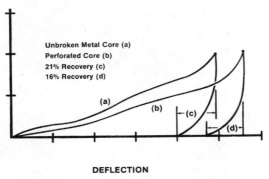

Figure 5.16 Load deflection and recovery.

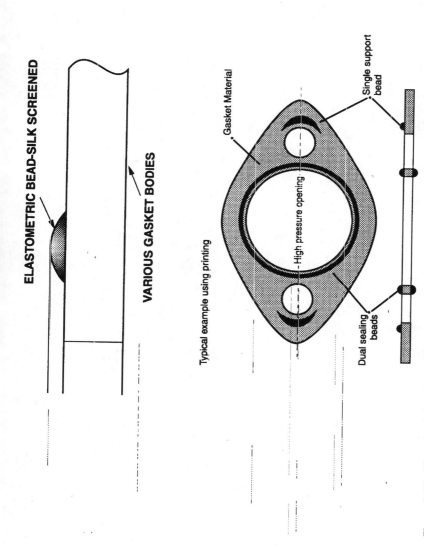

Figure 5.17 Elastomeric beads printed on a gasket body.

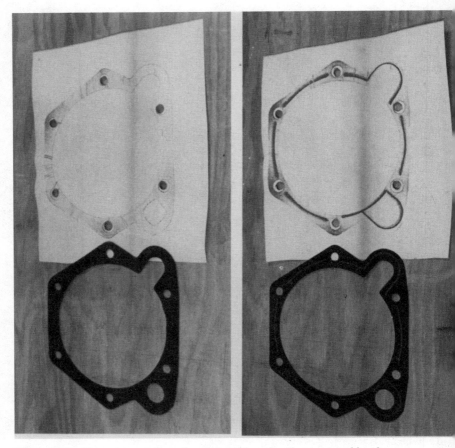

Figure 5.18 Stress impressions of gaskets with and without printed beads.

When the elastomeric beads are utilized on the gasket, higher resulting stresses of the elastomeric beads occur. This results in improved recovery properties of the gasket and improved long-term sealing. Figure 5.18 shows the stress distribution of gaskets with and without the printed beads. The improvement in stress distribution due to the beads is clearly shown.

A number of elastomers are used in this technique, but silicone is the most popular. Table 5.7 depicts a number of the elastomeric compounds that have been used in the silk screening deposition (printing) of beads on gaskets.

TABLE 5.7 Typical Specifications and Compatibility of Various Elastomeric Compounds Used for Printed Beads

MATERIAL PROPERTIES	SILICONE	PVC	FLEXIBLE EPOXY	NITRILE	PVC SPONGE
Durometer	40-70	40-60	70-90	60-80	20-40
Tensile, psi	800	1,500	1,500	1,200	500-800
Elongation, %	100-200	100-200	50-100	100-300	100-200
Max. Service Temp.	+400°F	+180°F	+225°F	+250°F	+150°F
Min. Service Temp.	-67°F	-20°F	-20°F	-40°F	-20°F
Flexibility	Can be bent on itself	Can be bent on itself	Can be bent on itself	Can be bent on itself	Can be bent on itself
COMPATIBILITY					
Water	Excellent	Excellent	Excellent	Excellent	Excellent
Grease & Oils	Excellent	Excellent	Excellent	Excellent	Excellent
Fuels	Good	Excellent	Excellent	Good	Good
Dilute Acids	Good	Good	Excellent	Good	Good
Dilute Bases	Good	Good	Excellent	Good	Good
Steam	Good	Not Resistant	Excellent	Good	Not Resistant

Incorporation of printed beads results in several benefits:

- Extreme flexibility and customized applications can be obtained with minimum tooling costs.
- The selective locations of the sealant bead allow for improved sealing by giving high unit loadings where required.
- Redistribution of loads as a result of sealant bead selection compensates for weak or distorted flanges.
- High-temperature resistance at specific locations can be realized with small amounts of sealant beads.
- Printed beads can be used successfully to modify and improve gaskets that have presented perennial sealing problems—and inexpensively.

- A wide range of servicing is available depending upon the material and type of gasket, the sealant bead material, and its thickness and shape.

In addition to the elastomeric sealing beads, rigid beads, such as those that result when rigid epoxies are utilized, are also incorporated sometimes into the gaskets. Their incorporation is for compression-limiting purposes and/or to reduce the flange distortion. In some gaskets, both the elastomeric bead and the rigid bead are utilized (Fig. 5.19). The deposition of elastomeric beads via silk screening is very common throughout the gasketing industry, particularly in internal combustion engine gasketing.

An extension of the silk screening technique is to deposit the beads using a tracer or robot. When this is done, it permits utilization of more material and increases the thickness possibilities for the beads. A further extension is to incorporate the elastomeric beads with an embossed material. In these cases, the elastomeric bead is deposited into the emboss and the combination can be likened to an O-ring being trapped. This results in an end product having very extensive recovery characteristics.

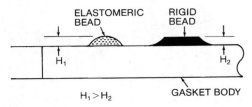

Figure 5.19 Elastomeric and rigid beads on a gasket body.

Screen manufacture

1. A synthetic mesh fabric is tightly stretched and fastened over a wooded or metal framework.
2. The screen fabric is coated with a photoreactive emulsion which is dried.
3. An opaque mask, in the configuration of the desired elastomer bead, is placed against the screen fabric.

4. The screen is exposed to a strong light source. The emulsion cures in areas not protected from the light source and becomes insoluble.
5. The emulsion remains uncured in the areas protected by the opaque mask and is rinsed from the screen with water or solvent.

Printing the bead

1. The screen is installed onto the printing machine and registered to the gasket to be printed so the bead will be deposited in the proper location.
2. A small quantity of liquid elastomer is poured onto the screen. A "flood bar" draws a uniform layer of the elastomer across the screen surface.
3. The screen is lowered onto the gasket. A squeegee is drawn across the surface, forcing the elastomer through the open bead pattern onto the gasket.
4. The printed gasket undergoes a curing operation to cross-link the elastomeric bead.

Another variation of the silk screening process is called mold-in-place (MIP). In this technique, the elastomeric bead is molded to the gasket. This results in a precision thickness variation of the bead and can result in a three-dimensional gasket. Some of the later designs incorporate liquid injection-molded silicone for this purpose.

Grommeting

Grommets in the gasket industry are rubber parts and/or rubber parts which are reinforced with metal or plastic. They are molded products and are added to the gasket to provide improved sealing at difficult-to-seal locations. Their cross sections are virtually unlimited and therefore permit a large range of design possibilities. Many materials are used for grommets. Most common are nitrile, neoprene, polyacrylic, silicone, and fluoroelastomer. Figure 5.20 depicts a few grommet designs.

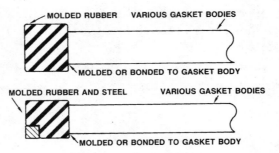
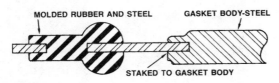

Figure 5.20 Various grommet designs.

The grommets are used in fluid passages where conformity with sealing surfaces and permeability are major problems and high fluid pressures are encountered. They have low spring rates and must be designed to have appropriate contact areas and restraint in order to effect high unit sealing stresses for withstanding the internal pressures. The grommets also have high recovery, which allows them to follow high thermal distortions associated with the application. Compression set and heat-aging characteristics must also be considered when elastomeric grommets are used.

Segmented Gaskets

Flat gaskets made up of several individual pieces must not be of a lower quality than gaskets made in one piece. Gaskets cannot be made in one piece when

1. The gasket dimensions exceed the production width of the material from which it is made or the gasket is larger than the machine holding fixture or other production facility.
2. The gasket price can be reduced as a result of saving material.

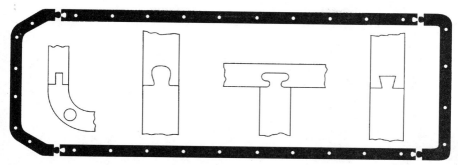

Figure 5.21 Various groove-and-tongue joints.

3. Reasons of assembly or packing and shipping make a one-piece gasket undesirable.

Large gaskets are blanked in one piece up to the full material width because of the high usefulness of the waste pieces. This is true except for very valuable materials, where it is less expensive to make up the gasket ring from several segments. Very large gaskets can be made in segments and can be folded for transportation.

The joint in the gasket should in principle be selected on the basis of optimum use of material (unless installation reasons do not permit); in any case it should not be located on bolt holes where the section is weakened. Figure 5.21 depicts various groove-and-tongue joints with either square, rounded, T-shaped, or dovetail-shaped tongues. Square groove-and-tongue joints are easy to fit but are also liable to creating a passage for the agent to be sealed off to escape. The dovetail joints, which are blanked with the aid of cutting tools, safeguard the sealing effect of the gasket.

References

Bickford, John H.: *An Introduction to the Design and Behavior of Bolted Joints,* Marcel Dekker, New York, 1990.
Czernik, D. E.: "Recent Developments and New Approaches in Mechanical and Chemical Gasketing," SAE paper 810367, February 1981.
Faires, V. M.: *Design of Machine Elements,* Macmillan, New York, 1955.
Rothbart, H. A.: *Mechanical Design and Systems Handbook,* 2d ed., McGraw-Hill, New York, 1985, sec. 27.4.
Standard Handbook of Machine Design, McGraw-Hill, New York, 1986, chap. 26, part 1, 1986.

Chapter 6

Rubber Gaskets

Rubber

Introduction

Rubber is a unique material. It is capable of stretching or compressing a large amount and rapidly recovering to its original thickness upon release of the deforming force. Rubber has memory, and the energy used to deform it is stored within the material. Deformation may be in compression or tension. The ability to undergo large deflection and store energy coupled with imperviousness make rubber an excellent gasket material.

The problem of selection and application of various rubber-like elastomers is difficult for many gasket engineers. Familiar engineering terminology, when applied to natural or synthetic rubbers, often means something other than the conventional usage of the word. Some design problems are a result of confusion stemming from a basic communications problem. In order to apply rubber, it is necessary that the characteristics of the material be understood. Rubber is a unique material and, unlike metals, often reacts in a completely different manner from what our experience with metals would indicate it should. For example, rubber displaces; it does not compress. All rubber properties are strain rate– and temperature-dependent.

Properties

Compression

As noted, rubber displaces and does not compress. For purposes of this handbook, rubber displacement is called rubber compression. Therefore, during application of compressive load, significant shear forces are developed. These forces can result in fracture of the rubber, and therefore limitations are imposed on the amount of compression. These limitations are different for the various elastomers.

Rubber is a vulcanized or cured product. The degree of cure significantly affects the material properties. Examples of the difference in the properties of uncured and cured rubber products are:

Uncured	Cured
Low strength	High strength
Poor recovery after stretching	Good recovery after stretching
High plasticity	Low plasticity
Soluble in solvents	Insoluble in solvents
High freezing point	Low freezing point
Low softening point	High softening point

Modulus

This property is reported as the force required to elongate the rubber a specified percentage, i.e., modulus at 100 percent elongation or modulus at 200 percent. This is a valuable process control number, and coupled with hardness gives a refined control on gasket deflection.

In rubber, as opposed to steel, the stress-strain relationship is not proportional. Therefore, the modulus of an elastomer refers only to a single point on the stress curve.

The procedure for conducting tensile tests is specified in ASTM D412. The usefulness of tensile tests to the rubber technologies is a means of control of this product. The design engineer, however, will have to decide if particular elastomer requirements will involve loading in tension. Elastomers are available with high tensile strengths, but as opposed to the practice in metals, tensile strength in rubber does not necessarily have any relation to other important design characteristics.

Hardness

Durometer reading is a measure of the hardness of rubber. It is associated with the rigidity or deformability of the rubber. It is measured on an arbitrary scale called the Shore A and covers a range of 0 to 100 degrees. For small deformation, it is directly related to Young's modulus, the force required to compress material. It is also related to shear modulus.

From a gasket design standpoint, it is one tool to control compression. If the flange is badly distorted, reducing the hardness will increase the gasket compression. It may be a more cost-effective way than increasing gasket thickness.

The procedure of hardness testing can be found in ASTM D676, D314, and D531. Hardness can be significant when it relates to the application of a material. It must be recognized that hardness readings are approximate. Also, hardness, as measured from small deformations, should not be interpreted as a measure of large deformations (stiffness).

Tensile and elongation

These properties are measured at the breaking point using the original thickness. They are very important values in controlling the rubber manufacturing process, but of little design value.

Compression set and creep relaxation

This property is determined by compressing the material a specific amount, then aging the compressed specimen at an elevated temperature for a specified length of time. If further curing takes place, the compression set will be higher than normal. If the compression is low or normal, the material is fully cured. State of cure or cross-linked density affects all rubber properties.

Resilience

This is the property of rubber to snap back to its original shape after having been deformed. It usually is expressed in percent of the energy that is returned after the removal of the stress that has caused a deformation.

Hysteresis

This represents energy lost per loading cycle. The stress-strain curve of a rubber compound plotted on extension and recovery shows an energy loss (100 percent minus the resilience percentage). This usually is manifested by the conversion of mechanical energy into heat.

Heat buildup

This is a term used to express temperature rise in a rubber product resulting from hysteresis caused by poor resilience of the rubber for the particular application.

The usual measure of resilience is not obtained by plotting a hysteresis curve but by measuring the rebound of a pendulum that has fallen from a given height. Another method measures the rebound of a metal ball as it is dropped on a test piece. Vibration methods also can be used. These tests are described in ASTM D1054, D945, and D623.

Resilience is significant to the designer as an indication of bounce in a piece of rubber. Low resilience, however, may be useful where certain damping characteristics are required. Heat buildup in the part, if excessive, is evidence of the need for a more resilient material.

Permanent set

This is the deformation of a rubber part that remains after a given time, when a specific load applied for a given time has been released. It is usual to specify whether the set has resulted from forces in tension, compression, or shear.

Standard tests for permanent set are described in ASTM D412, D395, and D1229. Set is expressed as a percentage of increase (when loaded in tension) or a percentage of original thickness (when loaded in compression).

Abrasion resistance

This is one of the remarkable properties of most rubbers. The term refers to the resistance of a rubber composition to wear and is measured by the loss of material when a rubber part is brought into contact with a moving abrasive surface. It is specified as percent of volume loss of sample as compared with a standard rubber composition.

Standard tests for abrasion are specified in ASTM D394 and D1630. Since abrasion tests do not duplicate service conditions, their usefulness is left to the interpreter. There are many types of abrasive actions and often elastomers that have provided excellent service in a particular application cannot be justified by conventional abrasive tests. A service test is recommended wherever possible.

Tear strength

This is a measure of the force required to propagate a cut in a normal direction to that of the applied stress, or the force required to initiate tearing. Standard test for tearing is described in ASTM D624. Tear resistance is affected by stress distribution, speed of stretching, and size of specimen. The tear resistance of rubber is greater at higher temperatures than at room temperature.

Permeability

This refers to the ease with which a liquid or a gas can pass through a rubber film. Standard tests for permeability are described in ASTM D814 and D815. Permeability is one of the most important characteristics of rubbers in the field design. Temperature is a very important factor, since the permeability at higher temperatures may be many times what it is at a low temperature. Also, permeability varies according to the compounding of the rubber rather than just according to type.

Flex fatigue

This is the result of rubber's being subjected to fluctuating stresses. Standard tests for flex fatigue are described in ASTM D430 and D813. The symptoms of flex fatigue commonly are used to compare elastomeric materials, especially unknowns, as compared with a rubber with an established performance.

Modulus of elasticity

This refers to the ratio of stress to strain. Since these values are not proportional, the modulus is not constant. Measurements made in tension, compression, or shear are valid at less than 15 percent strain because the curve is nearly straight in this portion of the stress-strain curve. Standard tests for stress-strain data are found in ASTM D575, D797, and D945.

Elastomers usually have two moduli of elasticity, a static and a dynamic. In resilient compounds, the dynamic modulus is only slightly greater than the static modulus. In low-resilience, high-hysteresis materials, the dynamic modulus can be much higher than the static modulus. This difference is also a function of temperature, compounding, and degree of vulcanization.

Rubber often is stressed in shear because greater deflections under a given load can be realized than when stressed in compression. Shear strain is defined as the ratio of linear deformation to the thickness of the rubber. Because of the many variables that affect the compounding and vulcanization of rubber, considerable tolerance must be allowed on rubber articles made to a specification. A ± 15 percent is normal, ± 10 percent is special, and ± 5 percent is fabulous.

While rubber can be deformed easily, it offers considerable resistance to changes in volume. This resistance usually is expressed as volume compressibility (a reciprocal of the bulk modulus) and is the reduction in volume per unit pressure. The volume compressibility of pure gum rubber at low pressures is about the same as that of water.

Shape factor

This refers to changes in the compression-deflection relationship because of the shape of the part. For pieces having parallel loading faces and sides normal to these faces, the ratio of the load face to the side area provides a measure of degree of compressibility. Since rubber is virtually incompressible, the ability of a rubber part to compress depends on the amount of side area that is free to bulge. Experimentation has shown that rubber specimens with the same shape factor and composition will deflect equally under a compressive stress.

Electrical properties

These properties of elastomers usually include insulation resistance, conductivity, dielectric constant, and power factor.

Insulation resistance is the resistance offered to the flow of current through an insulation under voltage stress. The property usually is considered only for dc potential and its value depends upon time of voltage application, stress value, and electric field configuration. Standard method of testing is by ASTM D257.

Conductivity of elastomers usually is considered where dissipation of static electricity is desired. The rubber compounds are considered conductive when they possess a direct current resistivity of less than 10^5 Ω-cm. ASTM D991 outlines a procedure for testing elastomers designed for electrical conductivity.

Dielectric strength measures an insulation's ability to withstand voltage. The rate of application, geometry of the electrodes, and shape of the test specimen all influence the results. Standard test methods are established under ASTM D149.

Dielectric constant or specific inductive capacitance compares the ability of an insulation to hold a charge with that of air. As such, it is expressed as a direct ratio of the test material to air. The dielectric constant is influenced by the power frequency. At 60 cycles the loss is usually small, but at high communications frequencies, the loss can become extremely serious. Dielectric constant and power factor tests are found in ASTM D150.

Adhesion

This refers to the strength of bond between a rubber and another material. It can be measured by applying a force to peel or strip the material, a force normal to the plane of the bond, or a force applied in shear. These are described in ASTM D413 and D429B for the first method, ASTM D429A for the second method, and ASTM D816 for the third. Adhesion does not depend entirely upon the elastomer, since metal or fabric when used also contribute to the strength of the bond.

Thermal properties

These properties of natural and synthetic rubber that are important include coefficient of expansion, thermal conductivity, and what is known as the "Joule effect." The coefficient of expansion of elastomers varies depending upon the kind and amount of filler added to the gum rubber. The more filler that is added, the lower the coefficient. The expansion of rubber is about once times that of steel. This means mold shrinkage is considerable and close tolerance on molded parts becomes quite costly. The best way to determine reasonable tolerances for a part is to consult the molder.

Thermal conductivity is the time rate of transfer of heat by the conduction for a unit thickness, over a unit area for a temperature differential. This property is important because it relates to the time necessary to heat the center of a molded part to vulcanizing temperature. The thermal conductivity of rubber is also important to the designer of products where heat is generated through their flexing, friction, or vibration and where provisions must be made for heat dissipation.

The "Joule effect" is a phenomenon based on the fact that the modulus of elasticity of rubber increases with a rise in temperature. If rubber is stretched and then heated, it tries to contract. Note that the rubber has to be under strain before this effect occurs. This apparent curiosity is important to the designer of many types of products where strain and heating can occur. Rubber torsion springs can be affected by stiffening as the temperature rises. Also, O-rings, if used to seal a rotating shaft, become heated and try to contract. This generates still more heat, eventually resulting in failure.

Low temperature

This affects rubber in several ways. Dropping the temperature very gradually stiffens the elastomer until a certain point is reached where stiffness increases a great deal for a slight decrease in temperature. Somewhere in this range, a point is reached where the specimen becomes brittle and will break or shatter on bending or impact. The point where brittleness occurs depends on the rate of application of load and seems to have no relation to the stiffness curve. The reaction is reversible, however, and an increase in temperature will restore the original properties of the rubber part.

Crystallization of rubber occurs at some moderately low temperature where the rubber molecules can move into a crystal pattern but do not have the energy to break loose again. As the crystal structure grows in size, the rubber becomes increasingly stiff. Brittleness is measured by ASTM D746; stiffness is measured by ASTM D676.

Weather and sunlight

These have an injurious effect on most rubber compounds. Sunlight causes the most rapid deterioration. Weathering causes natural rubber to crack because of ozone oxidation of the surface and a decrease in tensile strength and extendibility. Rubber technology, however, has provided types and compounds of rubber that will give good service and resistance when exposed to sunlight and weathering. Test methods using actual outdoor tests are described in ASTM D518 and D1171. Laboratory tests are described in ASTM D750.

Aging

This describes the deterioration of rubber with the passage of time. Storage deterioration was a problem years ago, but today compositions have been developed that provide rubber with excellent resistance to this type of degradation.

Many testing methods are used to test a rubber material for aging, most of which measure resistance to deterioration by oxygen and ozone. Some of these methods are described in ASTM D572, D454, D573, D865, D1206, and D1149.

Test results on aging are useful in making comparisons between different compounds and various types of rubber but may not be very helpful when the conditions of heat, light, air composition, or state of cure of the rubber are not known.

Rubber that is to be stored should be kept in a cool place, preferably in the dark, and should not be hung or coiled tightly or folded to place the rubber under strain. Stored rubber also should be kept away from ozone-generating equipment such as motors or switchgear.

Ozone deterioration

This results in tiny cracks at right angles to the direction of applied stress. It is a fairly common phenomenon, since ozone is generated in the air by sunlight and most rubber products are under stress during use.

Corona

This is the electrical discharge that takes place in the atmosphere around high voltage cables. Sometimes corona is visible as a violet glow. Corona causes the formation of ozone around the cable. Testing for ozone deterioration is specified in ASTM D1149 and D470.

Heat

This causes deterioration in natural and synthetic rubbers. The deterioration increases logarithmically with the temperature. Tests for heat deterioration are described in ASTM D4454, D573, and D856. Since small temperature changes can make large differences in degree of deterioration, a designer should be careful to select a realistic test temperature. Maximum operating temperatures may give a false indication of an elastomer's life expectancy when this temperature may be only rarely reached.

Water

This is absorbed by most natural and synthetic rubbers. Its effect, however, is much less than that of oil, and even at elevated temperatures, swelling from water is less pronounced. Tests for water resistance are given in ASTM D471. Results are reported as weight or volume change of the specimen.

Chemical resistance

This is a measure of the ability of the elastomer to be used in a chemical environment. There are no standard tests for this property since the conditions encountered are so variable. Laboratory tests are helpful in evaluating chemical resistance, but simulated service tests can be more useful.

Chemical resistance tables are published by most producers of natural and synthetic rubbers and should be consulted for specific information.

Oil deterioration

This is a result of the exposure of an elastomer to a liquid hydrocarbon. In some cases, this type of exposure will result in swelling of the material. Another effect can be the loss of the physical properties of the rubber. Standard tests for oil resistance can be found in ASTM D471. Proper selection of rubber type and rubber compound can provide oil resistance in applications where oil tolerance is of prime importance.

Other

Table 6.1 is a listing of the ASTM Test Methods for Vulcanized Elastomers, in this case, cured rubber gaskets.

Table 6.2 depicts the ASTM Suffix Letters and Test Methods for Vulcanized Elastomers. Again in this case it is applicable to cured rubber gaskets.

Manufacture

Rubber gaskets are either made from extruded sheet or molded into sheets or parts. In all cases, the rubber must be cured for the gasket with the required properties to be obtained.

After a rubber compound has been developed, the ingredients are mixed either by "open mill" mixing or internal mixing in a Banbury machine. The mixed materials are in a thermoplastic state and can be calendered into sheet, extruded into long continuous shapes, or molded. The first two are self-explanatory. Molded goods are produced by:

- Compression molding
- Transfer molding
- Injection molding

TABLE 6.1 ASTM Test Methods for Vulcanized Elastomers

D395	Compression Set Method "B"
D412	Elongation
D412	Tensile Module @ 100%, 200%, 300%
D412	Tensile Strength Die "C" (common)
D429	Adhesion Bond Strength Method "A"
D430	Flex Resistance
D471	Fluid Resistance Aqueous, Fuels, Oils, and Lubricants
D573	Heat Age or Heat Resistance
D575	Compression Deflection, Method "A"
D624	Tear Strength, Die "B" or "C"
D813	Crack Growth
D865	Deterioration by Heating
D945/D2632	Resilience
D1053	Low Temperature Torsional
D1171	Ozone or Weather Resistance
D1329	Low Temperature Retraction
D1418	Elastomer Classification
D2000/SAE J200	Common Elastomer Specification for Auto and Truck Components
D2137	Low Temperature Resistance "A"
D2240	Hardness Durometer "A"

In compression molding the mixed material is placed between halves of a mold, which is then placed and squeezed in a heated press for a specified period of time.

In transfer molding the mixed material is placed in a pot above the two mold halves. When the heated press closes, the material and a ram force pressure on the material, which is transferred from the pot into the mold through openings called sprues.

In injection molding the mixed material is in strip form. It is fed through an injection head into a heated, closed, and evacuated mold.

TABLE 6.2 Suffix Letters and Test Methods for Vulcanized Elastomers

Suffix letter	Test description	First suffix number (ASTM test method)							
		1	2	3	4	5	6	7	8
A	Heat resistance	D573 70 h	D865 70 h	D865 168 h	D573 168 h				
B	Compression set	D392 22 h solid	D395 70 h solid	D392 22 h piled	D395 20 h piled				
C	Ozone and weather resistance	D1171 ozone A	D1171 weather	D1171 ozone B					
D	Compression-deflection resistance	D575 A	D575 B						
EO	Fluid resistance (Coils and lubricants)	D471 oil no. 1 70 h	D471 oil no. 2 70 h	D471 oil no. 3 70 h	D471 oil no. 1 168 h	D471 oil no. 2 168 h	D471 oil no. 3 168 h	D471 fluid 101 70 h	D471 fluid designated fluid 70 h
EF	Fluid resistance (fuels)	D471 fuel A 70 h	D471 fuel B 70 h	D471 fuel C 70 h	D471 fuel D 70 h	D471 gasohol 70 h			
EA	Fluid resistance (aqueous)	D471 distilled water 70 h	D471 water and glycol 70 h						
F	Low-temperature resistance	D2137 A, 9.3.2 3 min	D1053	D2137 A, 9.3.2 2 h	D1329	D1329			
G	Tear resistance	D624 die B	D624 die C						
H	Flex resistance	D430A	D430B	D430C					
J	Abrasion resistance	To be specified							
K	Adhesion	D429 A	D429 B						
M	Flammability resistance	To be specified							
N	Impact resistance	To be specified							
P	Staining resistance	D429A	D429B						
R	Resilience	D945							
Z	Special, specified by user and supplier	To be specified							

Rubber gaskets

Design. Rubber gaskets can be plain die-cut parts from cured sheets or molded three-dimensional parts. In the latter, the parts can be molded to metals and/or plastics. Figure 6.1 depicts various molded gaskets.

Figure 6.2 depicts a metal core supported molded rubber gasket.

In some applications a compression limiter is needed to control the amount of compression on the rubber. Figure 6.3 depicts a proprietary design incorporating compression limiters.

Compression of the rubber can be controlled by edge molding the rubber to a rigid body. Figure 6.4 depicts an edge-molded design where the rigid body is metal.

Rubber compression can also be controlled by a volume control design. In these cases, metal and/or plastic containing a groove into which the rubber is molded is utilized. Figure 6.5 depicts this type of design.

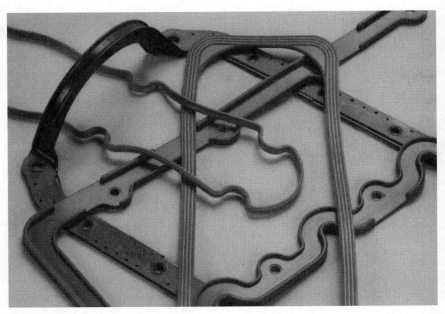

Figure 6.1 Various molded gaskets.

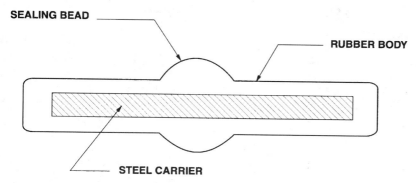

Figure 6.2 Metal core supported molded rubber gasket.

Properties. Rubber gaskets have a number of advantages compared to fiber backed gasket sheets. Some of these advantages are:

- Seals at low flange loads
- Three-dimensional design possible
- Excellent conformability to flange surfaces
- Complicated shapes available
- Excellent hysteresis properties
- Good recovery to follow dynamic flange movement
- Good noise isolation properties
- High reliability

One disadvantage is that molded rubber gaskets are more costly and require molds which also are costly.

Glossary of Terms Relating to Rubber and Rubberlike Materials

abrasion   the wearing away of a surface in service by mechanical action such as rubbing, scraping, or erosion.
accelerated test   (1) a test procedure in which conditions are intensified to reduce the time required to obtain a result. (2) Any set of test conditions designed to reproduce in a short time the deteriorating effect obtained under normal service conditions.
accelerator   a substance which hastens the vulcanization of an elastomer, causing it to take place in a shorter time or at a lower temperature.

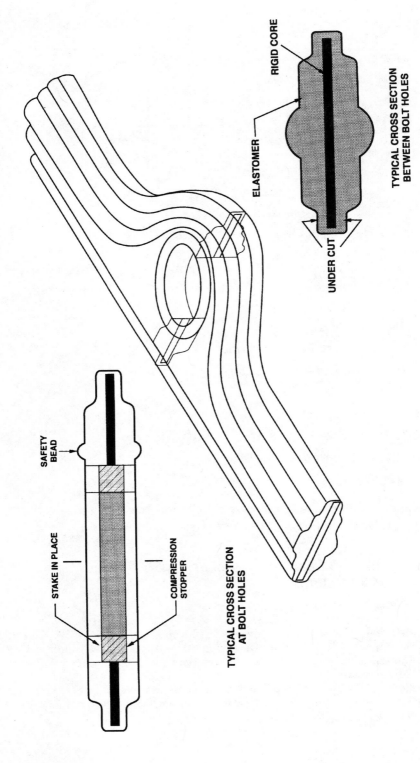

Figure 6.3 Compression limiter design.

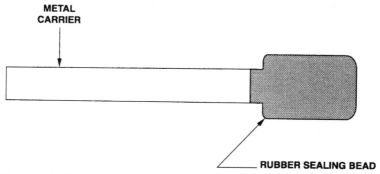

Figure 6.4 Metal carrier edge-molded to rubber gasket.

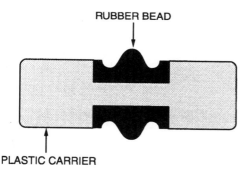

Figure 6.5 Volume control molded rubber gasket.

aging the changing of the material with time in response to its constitution and environment.
antioxidant an organic substance which inhibits or retards oxidation.
antiozonant a substance that retards or prevents the appearance of cracks from the action of ozone when the elastomer is exposed under tension, either statically or dynamically, to air containing ozone.
average room conditions 40 percent relative humidity at a temperature of 77°F.
benchmarks the lines on tensile specimens marking gauge length (q.v.).
brittle point the highest temperature at which an elastomer fractures in a prescribed impact test procedure. See freeze resistance and transition points.
calender a multiroll device used to form elastomer sheet to close tolerances or to build up an elastomer ply on a sheet of other material.
coefficient of elasticity the reciprocal of Young's modulus in a tension test.
cold flow a synonym for set or strain relaxation.
cold resistance a synonym for freeze resistance.
compounder's modulus a stiffness measurement extensively used by rubber technologies and expressed as "modulus at 300 percent" or "300 percent modulus" (any other percent elongation may be indicated, but 300 percent is

commonly used). By this is meant the tensile stress at the indicated elongation (see modulus).

compression modulus   the ratio of the compressive stress to the resulting compressive strain (the latter expressed as a fraction of the original height or thickness in the direction of the force). Compression modulus may be either static or dynamic.

compression set   the deformation which remains in rubber after it has been subjected to and released from a specific compressive stress or strain for a definite period of time at a prescribed temperature. Compression set measurements are for the purpose of evaluating creep and stress-relaxation properties of rubber.

concavity factor   rubber has no elastic limit, and the entire stress-strain curve is concave toward the stress axis or away from the strain axis. The relative amount that rubber varies from the Hooke's law ideal curve is known as "concavity factor" determined as the ratio between the energy of the extension curve to the straight-line curve and the same point. It may be expressed as

$$\frac{\text{Proof resilience}}{1/2 \text{ tensile product}}$$

conditioning   subjecting a material to a prescribed environmental and/or stress history prior to testing.

creep   the increase in deformation after initial deformation under sustained constant load.

crystallization   a change in physical properties resulting from crystalline orientation of molecular segments caused by temperature and stress.

cure   a synonym for vulcanization.

damping   the decay with time of the amplitude of free vibrations of a specimen (see hysteresis).

deformation   any change of form or shape in a body; in this book, the linear change of dimension of a body in a given direction produced by the action of external forces.

diluent   an inert powdered substance added to an elastomer to increase its volume.

drift   a synonym for set; also change in durometer reading with time.

dynamic fatigue   part failure under cyclic loading analogous to fatigue in metals testing.

dynamic modulus   the ratio of stress to strain under vibratory conditions. It is calculated from data obtained from either free or forced vibration tests, in shear, compression, or elongation. It is usually expressed in psi for unit strain.

dynamic resilience   the percentage of the vibrational energy which persists in the second of two successive free vibrations.

elongation   the deformation of the gauge length in a tensile test, expressed as a percentage of the original gauge length.

extruder   a screw-fed die which produces a continuous strip which can be of intricate cross section.

filler   a material added in substantial volume to alter end item properties.

flexing the repeated distortion of a material by bending, extending, or compressing forces or combinations of them.

flex resistance the fatigue life of material tested in a flexing machine according to a prescribed procedure.

freeze resistance the resistance to the effect of low temperatures of a specimen subjected to flexure, torsion, or impact according to prescribed test procedures. Freezing is a time-dependent phenomenon in that prolonged exposure to low temperatures results in increase in freezing point and brittle point as well as increased stiffness. See transition points.

freezing point the temperature at which a specimen exhibits a sudden increase in stiffness as it is slowly cooled. See freeze resistance and transition points.

gauge length (in this book) length over which deformation is measured.

growth the increase in dimension of an article resulting from a continued tensile stress being applied to the material during service.

hardness the resistance to surface indentation usually measured by the depth of penetration (or arbitrary units related to depth of penetration) of a blunt point under a given load using a particular instrument according to a prescribed procedure.

heat buildup the temperature rise in a part resulting from dissipation of applied strain energy as heat. See hysteresis.

heat resistance the property or ability of elastomers to resist the deteriorating effects of elevated temperatures.

hot tensile tensile strength at 212°F.

hysteresis the percent energy loss per cycle of deformation or the ratio of the energy absorbed to the total energy input.

immediate set the deformation found by measurement immediately after the removal of the load causing the deformation.

inert filler or pigment a material added in substantial volume to alter end item properties.

Joule effect the heating of an elastomer when it is elongated adiabatically and the contracting of elongated elastomers when heated.

memory the aftereffect response of a material to an event which, superimposed on the response to a later event, alters the behavior of the material from that which would result from response to a later event alone.

modulus (1) a coefficient of numerical measure of a property. (2) For elastomers, modulus usually refers to one of several measurements of stiffness or resistance to deformation. The use of the word without modifying terms may be confusing, and such use should not be encouraged. Modulus of elastomers may be either static or dynamic; static moduli are subdivided into tangent, chord, and compounder's. Compounder's modulus is always in tension, but all the others may be in shear, compression, or tension. Other terms used in connection with modulus are stiffness, rigidity, Young's, tangent, and elongation. (3) All elastic moduli in rubber (except compounder's) are ratios of stress to the strain produced by that stress; the strain is expressed fractionally; the units of the modulus are the same as those for the stress, usually psi.

modulus at 300 percent (or other percent elongation) the tensile stress at the indicated elongation. (Also see compounder's modulus.)

mold a form or matrix used to shape elastomers while in a fluid or plastic condition.

oil resistance the ability to withstand contact with an oil without deterioration of physical properties, or geometric change to a degree which would impair part performance.

permanent set (1) permanent set is the deformation remaining after a specimen has been stressed in tension a prescribed amount for a definite period and released for a definite period. (2) For creep tests, permanent set is the residual unrecoverable deformation after the load causing the creep has been removed for a substantial and definite period of time.

plasticizer a substance added to an elastomer to decrease stiffness or improve low-temperature properties, thus improving processing properties and altering end item properties.

proof resilience the tensile energy capacity of work required to stretch an elastomer from zero elongation to the breaking point expressed in foot-pounds per cubic inch of original dimension.

reinforcing pigment or agent a finely divided filler substance which, when properly dispersed in an elastomer, stiffens or produces improved physical properties in the vulcanized product.

relaxation decrease in stress under sustained constant strain or creep and rupture under constant load.

relaxation time the time required for a stress under a sustained constant strain to diminish to ($1/e$) of its initial value.

resilience the ratio of energy returned on recovery from deformation to the work input required to produce the deformation, usually expressed as a percentage.

rigidity a synonym for stiffness.

room temperature vulcanization (RTV) vulcanization by chemical reaction at room temperature.

set irrecoverable deformation or creep, usually measured by a prescribed test procedure and expressed as a percentage of the original dimension.

set at break elongation measured 10 min after rupture on reassembled tension specimen.

shape factor the distortional behavior of an elastomeric component depends upon exterior constraints, and for elastomers in compression this is accounted for by using an empirical function called shape factor.

shear modulus the ratio of the shear stress to the resulting shear strain (the latter expressed as a fraction of the original thickness of the rubber measured at right angles to the force). Shear modulus may be either static or dynamic.

shelf aging the change in materials properties which occurs in storage.

spring constant the number of pounds required to compress a specimen 1 in in a prescribed test procedure.

standard laboratory atmosphere whenever the materials to be tested are known to be sensitive to variations in temperature or moisture, or both, tests are conducted in a room or chamber of controlled humidity and temperature. Unless otherwise specified, the tests are made in the standard laboratory atmosphere having a relative humidity of 50 ± 2 percent at a temperature of $25 \pm 1°C$ ($73.4 \pm 1.8°F$).

average room conditions 40 percent relative humidity at a temperature of 77°F

dry room conditions 15 percent relative humidity at a temperature of 85°F

moist room conditions 75 percent relative humidity at a temperature of 77°F

static fatigue part failure under a continued static load, analogous to creep rupture failure in metals testing but often the result of aging accelerated by stress.

static modulus the ratio of stress to strain under static conditions. It is calculated from static stress-strain tests, in heat, compression, or tension. It is expressed in psi per unit strain.

stiffness the relationship of load and deformation. A term often used when the relationship of stress to strain does not conform to the definition of Young's modulus. See stress-strain.

strain a synonym for elongation. The term is used here as a synonym for deformation or to mean a deformation divided by gauge length.

strain relaxation synonym for creep.

stress a synonym for tensile stress. The term is used here to mean internal force per unit of area which resists change in size or shape of the body.

stress decay a synonym for stress relaxation.

stress relaxation the decrease in stress under sustained constant strain.

stress-strain synonym for stiffness, usually expressed in pounds per square inch or kilograms per square centimeter at a given strain.

subpermanent set strain retained at the end of a finite interval following release of stress.

T_{-10}, T_{-50} the temperatures at which a specimen which has been elongated, frozen, unloaded, and slowly heated recovers, respectively, one-tenth and one-half of initial elongation.

tangent modulus the slope of the line at any point on a static stress-strain curve expressed in psi per unit strain represents the tangent modulus at that point in shear, extension, or compression, as the case may be.

tear resistance the force required to tear completely across a notched specimen tested according to prescribed procedures expressed in pounds per inch of specimen thickness.

tensile product the product of tensile strength and elongation at break, usually divided by 10,000, or the product of tensile strength and gauge length plus deformation at break. The latter definition is an approximation (assuming no volumetric change) to actual rupture stress.

tensile pull failure load in a tension test.

tensile strength nominal stress at rupture (based on original cross-section area).

tensile stress a synonym for modulus.

tension modulus the ratio of the tension stress to the resulting tension strain (the later expressed as a fraction of the original length). Tension modulus may be either static or dynamic.

transition points the elastomeric behavior of certain high polymers is considered to be the phenomenological manifestation of thermal micro-brownian movements consisting of vibrations and rotations executed by molecular segments. These movements decrease with temperature so that a curve of, for example, stiffness versus temperature for rubber exhibits two transition points. The first is the temperature below which the elastomer is hard and stiff but not necessarily brittle. The other is the lowest temperature point which is called the glass transition point. The gradual disappearance of the rubberlike state and changes in other physical properties of this class of materials, as measured during cooling, is a function of a great many variables in the composition and service history of these complex materials a well as of test procedure. Thus the actual published values of transition temperature for the various elastomers furnish only a first approximation to the low-temperature limit of useful properties for elastomers. The actual service conditions must be considered in determining the effect of low environment temperature on component performance.

ultimate elongation   the elongation at rupture.

vulcanization   the process of creating a useful elastomeric material from basic compounding ingredients by heating the mixture with sulfur, or by other chemical reactions which result in change in material properties.

O-Ring Seals

Another design of molded rubber gaskets is called an O-ring. This product is a doughnut-shaped object that seals by being compressed between mating flanges which contain a groove in which the O-ring fits. This results in a metal-to-metal contact and very slight torque loss in the bolts. The groove is generally rectangular, and standards have been established for industrial as well as military applications. The military standard is more restrictive. The groove can either be in one face or one-half can be in either of the two faces (see Figs. 6.6, 6.7).

For most O-ring applications, a surface finish of 63 rms is recommended. In the cases of O-rings used for face seals in gas and vacuum applications, a surface finish of 16 rms has been

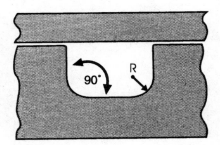

Figure 6.6   O-ring groove.

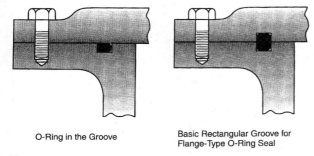

Figure 6.7 Regular and rectangular O-ring in grooves.

suggested. O-ring compression must be limited to avoid fracture, and the usual maximum limitation is set at 30 percent. The percent compression must take into account the tolerances of the grooves and the manufacturing tolerances of the O-rings, as they can significantly affect the percent of compression.

A computer program that does all the mundane percent compression calculations for various O-ring designs is "An Update of Design Guidelines for O-Ring Sealing Systems." It has been prepared by Robert L. Dega, Sealing Systems Consultant, 38002 Terra Mar, Mt. Clemens, Mich., (313) 468-5153. Figure 6.8 is an example of the output of this program.

MIL-P-25732C is an old military operation for O-rings. It covers O-rings for use with petroleum-based hydraulic fluids conforming to MIL-H-5606 and MIL-H-8382 over the temperature range of −65 to 275°F.

Military specification. As of Nov. 15, 1989, MIL-P-25732 is inactive for new design and is no longer used by the Department of Defense except in procurement of replenishment spares as necessary to maintain equipment in the field until obsolescence or wear-out.

The qualified products list (QPL) associated with this inactive for new design specification will be maintained until acquisition of the product is no longer required, whereupon the specification and QPL will be canceled.

Future designs should use MIL-P-83461, "Packing, Preformed, Petroleum Hydraulic Fluid Resistant, Improved Performance at 275°F."

212 Chapter Six

ORING GROOVE DESIGN AID				**ORING IN FLANGE FACE**	
ORING MEAN C.S.	Input In		0.070	**SECOND TRY**	
ORING SIZE-NUMBER	Input No		35	Input	0
ORING C.S. DIA MIN		In	0.067		0.067
ORING C.S. DIA MAX		In	0.073		0.073
ORING AREA MIN		In	0.00353		0.00353
ORING AREA MAX		In	0.00419		0.00419
ORING I.D. MIN		In	2.221		2.221
ORING I.D. MAX		In	2.257		2.257
GROOVE WIDTH MIN		In	0.094	Input	0
GROOVE WIDTH MAX		In	0.101	Input	0
GROOVE DEPTH MIN		In	0.051	Input	0
GROOVE DEPTH MAX		In	0.054	Input	0
GROOVE RADIUS	Input In		0.005	Input	0
GROOVE AREA MIN		In	0.00478		0.00000
GROOVE AREA MAX		In	0.00544		0.00000

SYSTEM PRESSURE**EXTERNAL**ORING CONTACTS GROOVE I.D.

GROOVE I.D. MIN	Input In		2.128	Input	0
GROOVE I.D. MAX	Input In		2.162	Input	0
ORING STRETCH MIN		%	-5.7		-100
ORING STRETCH MAX		%	-2.7		-100
GROOVE O.D. MIN (REF)		In	2.316		0
GROOVE O.D. MAX (REF)		In	2.364		0
ORING SQUEEZE MIN		%	19.4		100
ORING SQUEEZE MAX		%	30.1		100
VOLUME FILL MIN		%	64.9		#DIV/0!
VOLUME FILL MAX		%	87.7		#DIV/0!

SYSTEM PRESSURE**INTERNAL**ORING CONTACTS GROOVE O.D.

ORING O.D. MIN		In	2.355		2.355
ORING O.D. MAX		In	2.403		2.403
GROOVE O.D. MIN	Input In		2.330	Input	0
GROOVE O.D. MAX	Input In		2.350	Input	0
GROOVE I.D. MIN (REF)		In	2.128		0
GROOVE I.D. MAX (REF)		In	2.162		0
ORING O.D. COMPR. MIN		%	0.2		100
ORING O.D. COMPR. MAX		%	3.0		100

Figure 6.8 Computer output for an O-ring in the flange face.

SAE ARP 1234A

The Society of Automotive Engineers has an Aerospace Recommended Practice—ARP 1234A—which covers gland design, elastomeric O-ring seals, static axial, without backup rings. The purpose, scope, general requirements, and design criteria of this document follow.

Purpose. The purpose of this document is to provide the aerospace industry with standardized dimensional criteria for static axial elastomeric O-ring seal glands, without backup rings. It supplements ARP 1231, "Gland Design, Elastomeric O-Ring Seals, General Considerations."

Scope. This document establishes standard gland design criteria and dimensions for static axial O-ring seal applications without antiextrusion devices.

Gland configurations

General. A static axial O-ring seal is one which compresses the surfaces normal to the ID and OD of the O-ring. The most common axial gland configurations are depicted in Fig. 6.9. The preferred configuration is one that provides a complete groove in one part, as in Fig. 6.9.

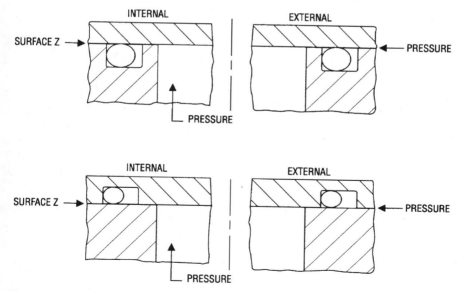

Figure 6.9 O-ring seal axial gland designs.

This configuration minimizes the number of gaps through which the seal can extrude and reduces the potential damage to the seal during assembly. The use of configurations with one gland wall missing, as in Fig. 6.10, is not recommended because it may result in damage of the seal through pressure reversals.

Noncircular glands. Although axial seals are usually circular, alternate configurations such as that in Fig. 6.11 are sometimes used. Placement of the seal within the groove depends on the direction of the pressure. Axial seals may be pressurized from a pressure source located within the seal's inner diameter (internal pressure application) or from a source located outside the outer diameter of the seal (external pressure application). The gland should be designed such that, prior to applying pressure, the seal will be in contact with the gland wall away from the pressure side of the seal.

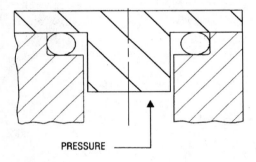

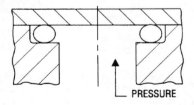

O-RING SEAL AXIAL GLANDS
NOT RECOMMENDED

Figure 6.10 O-ring seal axial glands not recommended.

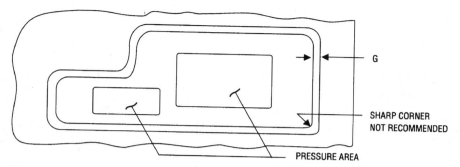

Figure 6.11 Example of a noncircular groove configuration.

Design criteria

Groove circumference or length

Internal pressure application. The groove circumference or length should be such that the outer diameter of the O-ring seal will be line-on-line with the outer wall of the groove when the seal is at nominal size and the groove outer wall is at maximum size.

External pressure application. The groove circumference or length should be such that the inner diameter of the O-ring seal will be line-on-line with the inner wall of the groove when the seal inner diameter is at nominal size and the groove inner wall is at minimum size.

O-Ring seal squeeze. To avoid leakage and to compensate for any sealing surface distortion during operation, the axial O-ring seal gland should be designed to achieve nominal squeezes as shown below.

O-Ring Seal Squeeze

AS 568 Dash no.	Nominal O-ring cross-section diam, in	Nominal squeeze, % of W nominal
−004 through −050	0.070	26
−102 through −178	0.103	24
−201 through −284	0.139	21
−309 through −395	0.210	18
−425 through −475	0.275	15

O-Ring seal swell. To accommodate O-ring seal swell, gland widths should be dimensioned to achieve a volume which is greater than the maximum swollen seal volume. In selecting gland widths, consideration must also be given to standardization of widths to cover a broad range of seal sizes. Such standardization is desirable to achieve reduction in inventory of tools and gauges.

The plug-to-bore diametral clearances shown below must be held to a minimum to prevent seal damage during operation. The recommended maximum clearance values follow.

Recommended Maximum Diametral Clearance

AS 568 Dash no.	Nominal seal cross section, in	Max diam clearance, in
−004 through −104	0.070	0.008
−015 through −050	0.070	0.010
−102 through −179	0.103	0.012
−201 through −284	0.139	0.016
−309 through −395	0.210	0.018
−425 through −475	0.275	0.024

O-rings are subject to extrusion in applications having high internal pressures. Figure 6.12 depicts the O-ring configuration while under increasingly higher pressures.

Extrusion is the deformation of a portion of the seal into the clearance between the mating metal parts. The extent of the deformation is determined by three main parameters:

1. The system pressure
2. The clearance or gap between the mating surfaces
3. The resistance of the seal material to deformation

It can be seen from Fig. 6.12 that the extrusion will be reduced if the system pressure is reduced. However, as the system pressure is usually defined, the design engineer must find other ways to overcome extrusion.

Another solution would be to reduce the extrusion gap. This can be accomplished by using tighter metal tolerances.

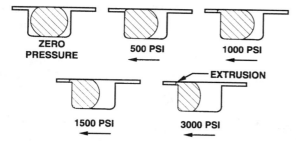

Figure 6.12 O-ring configuration under varying internal pressures.

Unfortunately this tends to sharply increase manufacturing and material costs and is therefore not considered a practical solution for most applications. In addition, the size of the extrusion gap will be affected by several other factors which cannot be readily controlled. These include:

1. Breathing of the metal parts under changing system pressure
2. Ovality between metal parts as is typically found in hydraulic cylinder applications
3. Wear and misalignment between the mating metal parts, a condition which unfortunately increases with time

Therefore, there is only one remaining parameter which can be practically manipulated to reduce the extrusion. This property is the resistance of the elastomeric material to deformation. The factors listed below will determine the ability of a sealing material to resist deformation and the resultant extrusion. The same factors will also have a significant effect on the service life of an O-ring. They are:

1. Hardness
2. Modulus
3. Dependence of the modulus upon temperature
4. Stress-relaxation characteristics of the material

Increasing the hardness and modulus will aid the O-ring to resist extrusion. Compounding the material so it will be little affected by increasing temperature will also aid in this regard. Finally, a compound with good stress-relaxation properties should be used.

There is a device which can be used for antiextrusion of O-rings. Called a backup ring, it is usually made from Teflon. Some of the Teflon backup ring characteristics are:

- Teflon backups have the extremely close tolerances so vital in precision applications. They offer the ultimate in reliability, uniformity, and reproducibility.
- Teflon, owing to its chemical inertness and homogeneity, may be used in any fluid. It will not embrittle, fibrate, or deteriorate under the most corrosive conditions.
- Unlike other materials, installation is easier because Teflon backups are scarf-cut or of spiral construction and slip easily over a shaft. Unlike leather, Teflon has no right or wrong side and requires no oil soaking or drying out to install.
- Teflon withstands high temperatures and can be used in 500°F continuous service.
- Low coefficient of friction helps reduce O-ring breakaway and spiral failure due to resin transfer.
- Teflon simplifies removal of old O-rings.
- Critical groove surface requirements are eliminated. This cost saving frequently pays for Teflon backup rings.

Figure 6.13 depicts the O-ring configurations under varying pressures with and without backup rings. The O-rings are molded from elastomeric compounds and are effective seals for

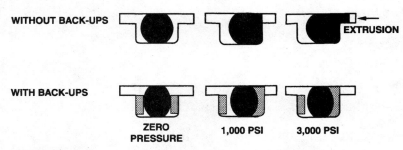

Figure 6.13 O-ring configuration with and without backup rings.

blocking the flow of liquids or gases. Careful consideration must be given to the size of the O-ring in regard to the size of the groove in which it is installed. Elastomeric O-ring materials are incompressible; that is, the volume remains the same during compression and only the shape of the O-ring is changed.

Minimum and maximum squeeze requirements are important. Percent squeeze is generally between 15 and 30 percent, although there are some variations on these percentages. O-ring manufacturers supply tables for groove dimensions that favor various diameters of O-rings so that the correct amount of compression of the O-ring results. Computer software is available for O-ring and groove design.

The grooves should have straight sides for reduction of the extrusion, but sloping sides at 5° are acceptable for pressures up to 1500 psi. To avoid unnecessary nicking or cutting of the O-rings during assembly, all the groove's corners must be rounded.

As with other types of seals, the first consideration for O-ring material selection is resistance to the media being sealed. Another consideration is the effect of pressure. Pressure tends to force the O-ring to either one or the other side of the gland depending on whether the pressure is internal or external. In some cases, backup rings made from harder materials are used.

The most common causes of O-ring failure are incompatibility of the elastomer and the media being sealed; incorrect O-ring size; improper O-ring groove design including excessively rough surface finish, and improper O-ring installation. In the case of improper elastomer selection, excessive compression can occur. There can be weather or ozone cracking. Heat hardening and oxidation need to be avoided. There can be extraction of the plasticizer.

If properly selected and designed, O-rings are very effective static seals.

References

Apple Rubber Products: "Seal Design Catalog," 1989.
Harper, C. A.: *Handbook of Plastics, Elastomers and Composites*, 2d ed., McGraw-Hill, New York, 1992.

Chapter

7

Gasket Testing

Gasket Material Tests and Their Significance

The table below depicts the identification, test method, and significance of various properties associated with gasket materials.

Property applications	Test method	Significance in gasket
Sealability	Fixtures per ASTM F37-89	Resistance to fluid passage
Heat resistance	Exposure testing at elevated temperatures*	Resistance to thermal degradation
Oil and water characteristics	ASTM F-104-93 and F146-93a	Resistance to fluids
Antistick characteristics	Fixture testing at elevated temperatures, including ASTM F607-90	Ability to release from flanges after use
Stress vs. compression spring rates	Various compression test machines	Sealing pressure at various compressions
Compressibility and recovery	ASTM F36-93 (sheet materials) ASTM F805-93 (composites)	Ability to follow deformation and deflection; indentation characteristics

Property applications	Test method	Significance in gasket
Creep relaxation and compression set	ASTM F38-93 and D395-59	Related to torque loss and subsequent loss of sealing pressure
Crush and extrusion characteristics	Compression test machines	Resistance to high loadings and extrusion characteristics at room and elevated temperatures

*Including high-temperature (900°F) creep relaxation.

Sealability

Sealability, of course, is the most important property of a gasket. Various companies have developed their own proprietary test procedures for determining gasket sealability. The universal test procedure is ASTM F37-89, Standard Test Methods for Sealability of Gasket Materials. The scope of this test procedure, referenced documents, summary of test methods, and significance and use follow.

Scope. The test methods provide a means of evaluating the sealing properties of sheet and solid form-in-place gasket materials at room temperature. Test Method A is restricted to liquid leakage measurements, whereas Test Method B may be used for both liquid and gas leakage measurements.

These test methods are suitable for evaluating the sealing characteristics of a gasket material under different compressive flange loads. The test method may be used as an acceptance test when the supplier and the purchaser have agreed to specific test conditions for the following parameters: test medium, internal pressure on medium, and flange load on gasket specimens.

This standard may involve hazardous materials, operations, and equipment. This standard does not purport to address all the safety problems associated with its use. It is the responsibility of the user of this standard to establish appropriate safety and health practices and determine the applicability of regulatory limitations prior to use.

Referenced documents. ASTM standards[1] D471, Test Method for Rubber Property—Effect of Liquids;[2] D2000, Classification System for Rubber Products in Automotive Applications;[3] E691, Practice for Conducting an Interlaboratory Study to Determine the Precision of a Test Method;[4] F38, Test Methods for Creep Relaxation of a Gasket Material;[5] F104, Classification System for Nonmetallic Gasket Materials.[5]

Summary of test methods. Both test methods utilize a test specimen compressed between the surfaces of two smooth steel flange faces. After the specified flange load is applied, the test medium is introduced into the center of the annular gasket compressed between the flanges and the specified pressure is applied to the medium. For liquid sealability tests (Test Methods A and B), reference fuel A (see Test Method D471, motor fuel section of annex) is recommended and the leakage rate is measured by a change in the level of a sight glass located in the line upstream from the gasket testing fixture. Nitrogen is the recommended gasket for the gas sealability test (Test Method B) and the leakage rate is measured by a change in the level of a water manometer located in the line upstream from the gasket test fixture.

Test Method A uses a test fixture by which an external load is transferred into the fixture to produce a compressive force on the gasket specimen.

Test Method B uses a test fixture in which the flanges are held within a four-bolt cage that permits loading the flanges at

---

[1]These test methods are under the jurisdiction of ASTM Committee F on Gaskets and are the direct responsibility of Subcommittee F-3.20 on Methods of Test for Nonmetallic Gaskets. Current edition approved July 28, 1989. Published September 1989. Originally designated as D2025. Redesignated in 1963 as F37-62T. Last previous edition F37-88.
[2]Annual Book of ASTM Standards, vol. 09.01.
[3]Annual Book of ASTM Standards, vols. 09.01 and 09.02.
[4]Annual Book of ASTM Standards, vols. 06.03 and 14.02.
[5]Annual Book of ASTM Standards, vol. 09.02.

various force levels. The flange load is measured by a transducer held within the cage.

Results of the sealability test are expressed as a leakage rate in milliliters per hour for the test specimen under the specific conditions of the test.

Significance and use. These test methods are designed to compare gasket materials under controlled conditions and to provide a precise measure of leakage rate.

These test methods are suitable for measuring leakage rates as high as 6 ml/h and as low as 0.3 ml/h. In many cases, "zero" leakage may not be attainable.

These test methods evaluate leakage rates after time periods that are typically 5 to 30 min under load. Holding a gasket material under load for extended time periods may give different results.

If the fluid being used in the test causes changes, such as swelling, in the gasket material, then unpredictable results may be obtained.

A new fixture has been developed to increase our knowledge of sealability of gasket materials. The fixture and the test results associated with it are described in SAE paper 900115, "Fluid Sealability of Gasket Materials—New Test Fixture, Instrumentation and Test Results." Figure 7.1 shows this fixture. Figure 7.2 is a close-up of the main test section. The abstract, design considerations, and summary from this paper follow.

Abstract. A gasket material fluid sealability test system has been developed using a fixture analogous to a bolted joint in an internal combustion engine. This system applies uniform load, simulates bolted joint rigidity, and measures leakage rate. It also measures load loss as a function of the material stress relaxation. It can be used with most liquids and gases over a broad pressure range.

Novel concepts are employed to improve fluid sealability testing, such as an innovative method of applying bearing load and electronic data acquisition that assures accuracy and increases operator productivity.

Gasket Testing 225

Figure 7.1 Sealability fixture.

Figure 7.2 Close-up of the main test section.

Design considerations. Most available fluid sealability test fixtures clamp the test specimen between two platens. Internal fluid pressure is applied and fluid leakage through and across the test material is measured over time.

From performing fluid sealability testing, current test equipment was found deficient in several aspects:

1. External loading equipment, such as a load frame, applies the required load to the test specimen.
2. Once loaded, there is no means of monitoring load loss due to fixture and test specimen settling.
3. Fixture stiffness cannot be changed.

To correct these deficiencies and to allow future improvements, a unique self-contained fluid sealability test fixture system was developed.

1. Using the wedge principle, the specimen is loaded uniformly and repeatably without an external load device.
2. Electronic sensing continuously records liquid level and load.
3. Adjustable fixture stiffness simulates a variety of bolted joints.

Summary. This fluid sealability test fixture system meets the following design criteria:

- A stand-alone system which contains a loading device and load monitoring instrumentation
- Simulates bolted joint rigidity
- Measures gasket material leakage rates
- Computerized data acquisition system
- Expandable for future improvements

This method is being evaluated by various companies, and ASTM is investigating it for possible incorporation as a new standard sealability procedure.

There are other standards associated with the sealability of the gasket. The table on p. 228 gives a survey of existing standard test methods for the measurement of sealability.

Heat resistance

Gasket materials must be resistant to the temperature of the application. Thermal degradation is associated with all gasket materials. Some materials can withstand considerably higher temperatures than can others. In some cases the thermal degradation can result in significant leakage while in other cases it poses no problem. Gasket manufacturers should be consulted in this regard.

Various companies have developed their own fixtures and test procedures to determine the heat resistance of gasket materials. In general they consist of clamping the material to the compressive stress of the application and then placing the fixture in an oven at or slightly above the temperature of the application. Exposure time varies with companies.

Fluid resistance

Resistance to the media being sealed is another important property of a gasket material. Consultation with gasket suppliers should be done if there is any doubt about a particular gasket's resistance to a specific fluid.

For generic purposes, ASTM has developed oil and water immersion test procedures to aid in determining resistance to fluids. This procedure is ASTM F146, Standard Test Methods for Fluid Resistance of Gasket Materials.

The fluids used in this immersion testing are ASTM oil numbers 1 and 3, fuel B, distilled water, or other test fluids as needed. The scope, referenced documents, summary of test methods, and significance and use follow:

The fact that there are at least five standard test methods is an indication that none of the standards meet all the requirements of the producer and the user. ASTM F37 employs a very narrow width of test ring and a very low internal pressure.

ASTM F112, developed for sealability tests of enveloped gaskets, may be used for other sealing materials as well. The nec-

Standard Test Methods for the Measurement of Sealability

Test method	Gasket dimensions, in	Gasket stress, psi	Internal pressure, psi	Test fluid	Leak detection
ASTM F37, sealability test	1.74 × 1.27 × 0.03	125–4000	max. 14.5 (760 mmHg)	ASTM Fuel A	ml/hour penetrated fluid
ASTM F112, sealability of enveloped gaskets	6.5 × 4	Variable	90	Air	Pressure drop after 24 h
ASTM F401, determination of gasket factors	6.19 × 4.5 (ANSI B16.5 class 300 4-in flanges)	Variable	2–300	a. Refrig. R12 b. Water	Halogen 1 detector Pressure drop
ASTM F586, leakage rates vs. y and m	6.19 × 4.5 (ANSI B16.5 class 600 4-in flanges)	Up to 6770	2–300	a. Nitrogen b. Water	ml/s penetrated gas or water
DIN 3535/4	3.54 × 1.96	4350 (hydraulic press)	580	Nitrogen	ml/min penetrated gas

essary gasket load to seal off an air pressure of 90 psi is measured. The test requires 24 h for one reading.

ASTM F401 was intended to determine gasket factors by varying load and internal pressure. A standard 4-in flange connection is used. Whereas gasket load with the help of strain-gauged bolts is determined with high accuracy, the leakage measurement, using a halogen detector, is not quantitative and allows for a yes-no leakage determination.

ASTM F586, again intended to establish values for y and m factors, is using the heavier class 600, r'' flanges to avoid flange bending. Gasket load, internal pressure, and leakage are measured quantitatively, and therefore a means is provided to determine leakage under controlled conditions.

DIN 3535/4 is a German standard for the measurement of a gas leakage and is widely used in Europe to determine leakage rates at standard load and internal pressure.

Since it is intended to characterize gasket properties under idealized conditions, the gasket load is not applied with bolts but with a hydraulic press. This ensures uniform load distribution all over the gasket area. The surface of the pressure platens is polished and the gasket surfaces are covered with a thin plastic foil, thereby eliminating surface leakage.

Scope. These test methods cover the determination of the effect on physical properties of nonmetallic gasketing materials after immersion in test fluids. The types of materials covered are those containing asbestos and other inorganic fibers (type 1), cork (type 2), and cellulose or other organic methods are not applicable to the testing of vulcanized rubber, a procedure that is described in Test Methods D471. It is designed for testing specimens cut from gasketing materials or from finished articles of commerce.

This standard may involve hazardous materials, operations, and equipment. This standard does not purport to address all the safety problems associated with its use. It is the responsibility of the user of this standard to establish appropriate safety and health practices and determine the applicability of regulatory limitations prior to use.

Referenced documents. ASTM standards D412, Test Methods for Rubber Properties in Tension; D471, Test Method for Rubber

Property—Effect of Liquids; F36, Test Method of Compressibility and Recovery of Gasket Materials; F104, Classification System for Nonmetallic Gasket Materials; F147, Test Method for Flexibility of Nonmetallic Gasket Materials; F152, Test Methods for Tension Testing of Nonmetallic Gasket Materials.

Summary of test methods. Appropriate test specimens are subjected to complete immersion in test fluids. After the specimens are immersed in the various test fluids, the effect on physical properties is expressed as change in tensile strength, compressibility in softened condition, flexibility, volume change, and thickness and weight changes from the original condition.

Significance and use. The test methods provide a standardized procedure to measure the effect of immersion in specified fluids under definite conditions of time and temperature. The results of this test are not intended to give any direct correlation with service conditions in view of the wide variations in temperature and special uses encountered in gasket applications. The specific test fluids and test conditions outlined were selected as typical for purposes of comparing different materials and can be used as a routine test when agreed upon between the purchaser and manufacturer.

Antistick characteristics

The gasket's ability to release from the mating flanges after use is important in some applications. Scraping of the gasket can cause damage to the mating flanges and gasket particles can fall into the sealed media, creating potential problems. Various test procedures have been developed by gasket companies to determine antistick properties of different gasket materials under different environments. Again, they usually involve clamping the material between appropriate mating flanges at the stress level of the application. The fixture then is immersed or subjected to fluid passage for a period of time. Disassembly forces can be qualitatively or quantitatively determined. Adhesion is usually rated by visual observation.

ASTM has a test procedure: ASTMs 607 Test Method for Adhesion of Gasket Materials to Metal Surfaces. The scope, referenced documents, summary of method, and significance and use follow.

Scope. This test method provides a means of determining the degree to which gasket materials under compressive load adhere to metal surfaces.

The test method may be employed for the determination of adhesion. The test conditions described are indicative of those frequently encountered in gasket applications. Test conditions may also be modified in accordance with the needs of specific applications as agreed upon between the user and the producer.

This standard may involve hazardous materials, operations, and equipment. This standard does not purport to address all the safety problems associated with its use. It is the responsibility of whoever uses this standard to consult and establish appropriate safety and health practices and determine the applicability of regulatory limitations prior to use.

Referenced documents. ASTM standards F38, Test Methods for Creep Relaxation of a Gasket Material; F104, Classification System for Nonmetallic Gasket Materials.

Summary of method. The procedure described in this test method involves placing a specimen of a gasket material between flat platens of the desired metal, loading the assembly, and subjecting it to a specified set of conditions. The tensile force required to part the platen is measured.

Significance and use. This test method provides terms such as megapascals or pounds-force per square inch of gasket surface for expressing the extent of adhesion applicable to all materials within the scope of Classification System F104. Under certain conditions, adhesion develops when gasket materials are confined in a compressed state between metal flanges. Adhesion is important as an index of ease of removal of a gasket material. Since other variables may enter into the performance in an application, the results obtained should be correlated with field results. A typical set of conditions is described in this test method. This test method may be used as an acceptance test when agreed upon between the user and the producer.

Stress vs. compression spring rates

As noted earlier, the stress versus compression for gasket designs is important, as adequate stress at various sealing locations is necessary. In addition, adequate compression to compensate for flange unflatnesses must be available. The stress-compression graphs are usually obtained via use of compression test machines. Both static and dynamic spring rates are determined during this type of testing.

Compressibility and recovery

The compressibility of gasket materials is important, as noted above. The recovery of the gasket is important in that this property determines if a material can follow the deflections that occur during unit operation and warpage of the flanges that occurs with time of use.

ASTM has a test procedure: A36, Test Method for Compressibility and Recovery of Gasket Materials. The scope and referenced documents of this procedure follow.

Scope. This test method covers determination of the short-time compressibility and recovery at room temperature of sheet gasket materials and in certain cases gasket cut from sheets. It is not intended as a test for compressibility under prolonged stress application, generally referred to as "creep," or for recovery following such prolonged stress application, the inverse of which is generally referred to "compression set." Also, it is not intended for tests at other than room temperature.

Referenced documents. ASTM standards E691, Practice for Conduction of an Interlaboratory Study to Determine the Precision of a Test Method; F104, Classification System for Nonmetallic Gasket Materials.

Creep relaxation and compression set

Creep relaxation and its relationship to torque loss have been discussed earlier. Various test methods for determining the relaxation of gasket materials have been developed by makers and users of gaskets. ASTM has a test procedure for this property. It is ASTM F38, Standard Test Methods for Creep Relaxation of a Gasket Material. The scope, referenced documents, summary of test methods, and significance and use follow.

Scope. These test methods provide a means of measuring the amount of creep relaxation of a gasket material at a stated time after a compressive stress has been applied.

Test Method A, creep relaxation measured by means of a calibrated strain gauge on a bolt.

Test Method B, creep relaxation measured by means of a calibrated bolt with dial indicator.

The values stated in SI units are to be regarded as a standard.

Referenced documents. ASTM standards D3040, Practice for Preparing Precision Statements for Standards Related to Rubber and Rubber Testing; F104, Classification System for Nonmetallic Gasket Materials. ASTM adjuncts: Relaxometer, Method A (F38); Relaxometer, Method B (F38).

Summary of test methods. In both test methods the specimen is subjected to a compressive stress between two platens, with the stress applied by a nut and a bolt.

In Test Method A, normally run at room temperature, the stress is measured by a calibrated strain gauge on the bolt. In running the test, strain indicator readings are taken at intervals of time, beginning at the loading time, to the end of the test. The strain indicator readings are converted to percentages of the initial stress which are then plotted against the log of time in hours. The percentage of initial stress loss or relaxation can be read off the curve at any give time, within the limits of the total test time.

In Test Method B, run at room or elevated temperatures, the stress is determined by measuring the change in length of the calibrated bolt with a dial indicator. The bolt length is measured at the beginning of the test and at the end of the test; from this the percentage of relaxation is calculated.

Significance and use. These test methods are designed to compare related materials under controlled conditions and their ability to maintain a given compressive stress as a function of time. A portion of the torque loss on the bolted flange is a result of creep relaxation. Torque loss can also be caused by elongation of the bolts, distortion of the flanges, and vibration; therefore, the results obtained should be correlated with field results. These test methods may be used as a routine test when agreed upon by the consumer and the producer.

Compression set is determined by measuring a gasket's thickness, clamping it in a fixture, subjecting the assembly to the specified test conditions, disassembling the fixture, and measuring the recovered thickness of the gasket.

Compression set by definition is

$$\frac{T_o - T_f}{T_o} \times 100 = \text{percent compression set}$$

where T_o = original thickness of material
T_f = final thickness of material

The amount of compressive set is directly related to the torque loss of the application. The loss in thickness of the gasket from its initial compressed thickness to its thickness during use is directly related to the loss in stretch of the clamping fasteners. The test for compression set is related to this amount. A better test is to use lead pellets to measure the internal clamped gasket thickness and also the final clamped gasket thickness after unit operation. This requires two tests: (1) a clamp up and disassembly and (2) a reassembly and unit operation with the installed pellets.

Crush and extrusion characteristics

In some applications, gaskets and certain locations on gaskets may be subjected to clamping stresses of sufficient magnitude to cause extrusion (compressive yield). Testing for crush and extrusion utilize compression test machines and both room and elevated temperatures. A quantitative method used in this determination is to measure the area of the test specimen before and after testing and compare the areas. Some of the test procedures include an immersion of the materials in a fluid similar to or in the fluid to be sealed, prior to the application of the potential extruding load.

Another test device used to measure crush resistance of a gasket material utilizes a fixture to determine split resistance of the material. The split resistance test is used to evaluate the performance of gasket material resistance to splitting when exposed to high localized compressive stress, sealed medium, and elevated temperature.

This test is accomplished using small bolted-together-type plate fixtures, in combination with a small-diameter steel wire to apply localized stress to the gasket material. In addition, high-temperature and sealed-medium conditions are simulated by presoaking the material (before stress application in the test fixture) in the sealed medium and then heat soaking the material in an oven. Figure 7.3 depicts the split resistance fixture.

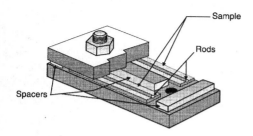

Figure 7.3 Split resistance fixture.

Gasket Material Analysis Techniques

Thermal analysis

A number of analytical instruments are used to identify and screen potential materials for use in various gasketed joints. The information can be used to identify components, detect variations in formulations, and infer performance of the final product. The analysis involves techniques for characterizing materials by measuring changes in physical or chemical properties resulting from controlled changes in temperature.

The various techniques for thermal analysis (TA) are as follows:

Differential scanning calorimetry (DSC)

Differential thermal analysis (DTA)

Dynamic mechanical analysis (DMA)

Thermal gravimetric analysis (TGA)

Thermal mechanical analysis (TMA)

TA instrumentation utilizes a furnace in which the sample is heated at a controlled rate in a controlled atmosphere and environment. A selective transducer monitors changes in the material by generating a voltage signal. After amplification and storage on magnetic disk along with temperature information, the data are recorded on a printer. Utilization of computers permits data analysis quickly.

Differential scanning calorimetry (DSC). This is probably the most widely used thermoanalysis technique. It is used to measure the heat flow absorbed or evolved of a material during phase change. DSC can measure both temperatures and heats of transition or reactions. DTA had been the primary TA technique and recently has been displaced with DSC.

The sample in a reference is placed in the pans, which rest on raised portions on a disk. Heat is transferred into the disk through the sample and the reference. The differential heat flow between the two is monitored by thermocouples. Some of the applications that DSC can perform include thermal transitions and polymers, crystallization, glass and melting transmissions, curing reactions, oxidative stability of lubricants and polymers, kinetics of thermal sets, purity of pharmaceuticals and organics, melt temperature, and catalyst efficiency.

Differential thermal analysis (DTA). A major application for DTA is a high-temperature characterization of minerals, alloys, ceramics, and glasses. Again, a sample and reference are located in the programmable furnace and thermocouples are used to measure temperatures. They measure both the presence of transitions and the temperatures at which they occur.

Dynamic mechanical analysis (DMA). DMA is used to measure mechanical properties as they are deformed under repeated stress. Modulus of a material as well as its stamping characteristics are primarily measured. Parameters define the inherent stiffness of the material and its tendency to convert mechanical energy into heat upon being stressed. A sample is rigidly clamped between two arms where the arms on the sample are part of a combined resonant system. The position of one arm in a pivot is fixed and the other arm in the pivot is movable by means of a mechanical slide. This permits a wide variety of sample lengths. A pair of arm locking pins is used to immobilize the arms and maintain them in parallel during sample mounting. When the pins are removed, the system is displaced and set into oscillation. As the system oscillates, the sample is subjected to fluctual deformation. Amplitude is measured by a linear variable differential transformer (LVDT). The frequency of the oscillation is related to the modulus of the sample, and the energy needed to maintain this amplitude is a measure of the damping of the sample. Some of the applications for DMA include curing of thermosets, polymer blend compatibility, impact stability with damping correlation, observation of a plasticizer fix, sound and vibration dissipation correlation, and characterization of supporting systems such as coatings and adhesives.

Thermal gravimetric analysis (TGA). This technique involves measurement of a weight change as a function of temperature. Samples are usually heated at some selected rate in an oxidative or inert atmosphere. An electrical mechanical transducer is used in this technique. A beam is maintained in the null position by regulation and amount of current flowing through the transducer coil. A pair of photosensitive diodes acts on the position sensor to detect movement of the beam.

The current flowing through the coil determines the weight signal that is supplied to the output circuitry. Some of the better techniques for TGA include thermal stability materials, kinetic studies, reactive atmosphere analysis, corrosion resistance and corrosion studies, volatiles and moisture determinations, accelerated aging tests, and oxidation-reduction reactions. The largest use of TGA is for gasket compositional determinations. Figure 7.4 depicts a typical TGA plot of a nonmetallic gasket material.

Thermal mechanical analysis (TMA). This is used to characterize the dimensional changes of material as a function of temperature. The changes can be linear or volumetric. TMA equipment includes a linear variable differential transformer and thermocouples in direct contact or in close proximity to the samples. The final output is a plot of the probe displacement versus either the sample's temperature or time.

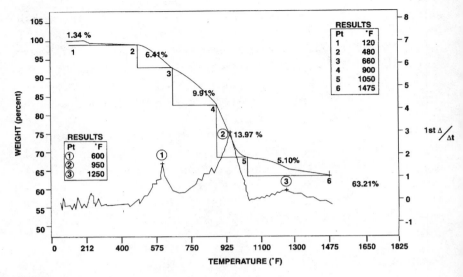

Figure 7.4 TGA plot of a nonmetallic gasket material.

The sheet tested was an aramid fiber sheet with a nitrile binder. Also present were cellulose fibers. Cellulose loss occurred around 300°F. Binder loss was at 250 to 500°F. Aramid loss was at 475 to 575°F.

Infrared spectroscopy

Infrared spectroscopy is the study of the interaction of infrared energy with matter. Most compounds will absorb infrared energy at specific wavelengths depending on the types of chemical bonds present. It's these different absorption tendencies that yield a characteristic "fingerprint" spectrum, making infrared analysis such a valuable tool.

Applications for infrared spectroscopy include:

1. Binder and fiber determinations of facing materials
2. Coating, adhesive identifications
3. Additive identifications, such as filler, plasticizer, antioxidant, and curatives
4. Polymer identification of rubber parts and bulk material
5. Anaerobic component analysis

6. Quality control applications for product uniformity confirmations

Chromatography (liquid and gas)

Chromatography is a separation technique whereby components are separated between two phases: a solid, stationary bed phase and a mobile phase moving through this bed. Separation occurs owing to the differences in absorption tendencies of the various materials in the sample to the stationary phase. This technique finds its greatest use in the quantitative analysis of components. The technique is also valuable qualitatively, to confirm component composition.

In liquid chromatography, various solvents are used to move a sample through the stationary phase and then are detected with an ultraviolet detector. Samples must be soluble in the LC mobile phase and free from any particulates. Significant added time is also required for methods development and standards quantification. The resultant chromatogram obtained is a plot showing absorption versus elution time of the components present.

Applications include many involving quantitative determinations, especially when dealing with complex multicomponent systems such as anaerobics, rubber compounds (uncured), coatings, and adhesives.

In gas chromatography, a sample is injected onto a column. Carrier gas is then used to propel it through as it is heated at a programmed rate. Separation occurs owing to the different vaporization temperatures of materials. The vapor is detected using one of two detectors: (1) a flame ionization detector and (2) a thermal conductivity detector. Samples must be volatile at up to 350°F and free from any particulates.

Applications are similar to those of liquid chromatography but are geared to the analysis of more volatile components.

Visual (microscopic) analysis

Two types of microscopes are normally used for visual analysis of gasket materials:

1. Regular light
2. Polarized light

Of the two, polarized light microscopy yields the most information. Besides the actual "visual" examination of material, the greatest use of polarized microscopy is in particle and fiber characterization. Polarized light is bundles of light rays with a single direction. This technique provides a wide variety of observations, including:

1. Color and transparency
2. Refractive index determinations
3. Birefringence (color visible owing to double refraction in crossed polars)
4. Pleochromism (color appearance caused by different vibrational directions with a single pole)

Scanning electron microscopy (SCM)

This instrumentation is very expensive and is not usually a part of a gasket company's equipment. Specialized laboratories that have SCMs are used for testing requiring this type of analysis. An electron beam is impacted on the test specimen and high-quality imaging and/or particle identification is made.

Gasket Material and Gasket Testing

Stress-compression (Instron)

A testing device normally used for stress-compression testing is the Instron machine (Fig. 7.5). The Instron is a mechanically actuated testing device which allows one to compress or stretch test materials to be evaluated. With the addition of an environmental chamber, the analysis of test materials can also be done in environments similar to those in actual operating conditions.

The Instron, equipped with two load cells and two sensitivity ranges, can accurately operate anywhere from 10 to 20,000 lb of force.

The environmental chamber can be accurately operated and controlled with a plus or minus tolerance of 1°F ranging from

Gasket Testing

Figure 7.5 Instron machine.

+ 600°F down to −100°F with the addition of a CO_2 attachment primarily used for deep thermal cycles.

The Instron can also be used in conjunction with a linear variable displacement transducer (LVDT). When the LVDT is installed and zeroed out on the Instron, machine deflection is ultimately eliminated.

Another feature of the Instron is the chart recorder. With the chart recorder, it is possible to control and monitor full-scale load of the Instron and stroke of the crosshead. The chart can also be used in conjunction with the LVDT.

The speed of the crosshead can also be controlled by interchangeable gears on the Instron drive mechanism.

Stress compression (MTS)

Another load-compression machine which extends the dynamic testing of the Instron is a material testing service (MTS) load frame (Fig. 7.6).

Figure 7.6 MTS machine.

The MTS load frame is a servocontrolled electrohydraulic system, which is powered by a hydraulic power supply. This material test system is used for either static or dynamic testing in either compression or tension with a maximum load of approximately 100,000 to 150,000 lb.

There are four different waveforms that can be generated, of which one is used for the cyclic testing of the specimen; they are a sine wave, haversine, haver square, or a ramp. There are three control modes, of which one is selected to control the test; they are:

- Load control
- Strain control
- Stroke control

The load frame also has its own built-in X-Y recorder. This is generally used for L/D curves but also can be used for short-term dynamic tests. The recorder may have load, strain, or stroke on either the X axis or the Y axis.

The MTS hydraulic supply can also be used as a pulsator.

Pulsator

The pulsator is actually three hydraulic flow control devices functioning as a single unit. It can be used to pressurize one or two enclosed volumes, such as a gasketed joint.

Some of the parameters tested are for gasket relaxation, eyelet cracking, dynamic motion, and flange brinelling.

By using a hot oil circulating system to simulate unit operating conditions, component temperature can be raised to the maximum temperature resistance of the oil. (See Fig. 7.7.)

Bolt stretch

When testing gaskets in the actual application, accurate data concerning the fastening bolt stretch are desired. A unit which can be used in this regard is an ultrasonic stretch measuring instrument (Figs. 7.8 and 7.9). This ultrasonic instrument is specifically designed for measuring the change in length of threaded fasteners. As fasteners are tightened in a joint, they

Figure 7.7 A pulsator.

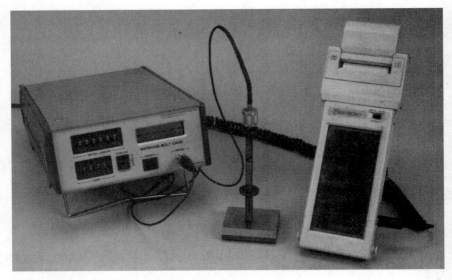

Figure 7.8 Ultrasonic instrument using digital readout.

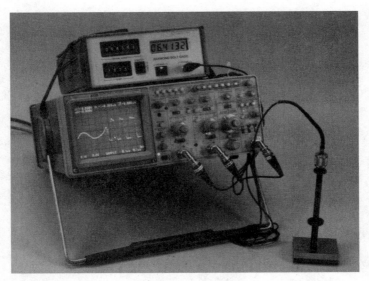

Figure 7.9 Ultrasonic instrument utilizing digital and CRT readouts.

stretch slightly, and the amount of that elongation is directly related to the amount of clamping.

By accurately measuring the stretch, the instrument may then be used at a later time to check for relaxation or loss of tension (stretch).

It operates by transmitting an ultrasonic pulse into one end of a fastener by means of a transducer. The pulse travels through the bolt to the opposite end and is reflected back. The time it takes for the pulse to travel from the transducer and back again is translated electronically into a measurement of length.

Comparisons of measurements before and after loading provide an indication of the fasteners' stretch (and/or loss of stretch). The unit measures stretch to an accuracy of ± 0.0001 in. Various parameters, however, must be taken into consideration; these are temperature of the material and the material's stress factor.

Some advantages in reference to gasket application:

1. Control bolt preload by ultrasonically measuring elongation.
2. Control of both dynamic and static measurements can be made. This means that one can monitor short- and long-term relaxation effects, external load effects, temperature effects, vibration loosening, and fatigue loading.

Surface finish

Surface texture is the repetitive or random deviations, from the nominal, which form the three-dimensional topography of the surface. There are four features that constitute surface texture: roughness, waviness, flaws, and lay.

Roughness is the irregularities in the surface texture which are inherent in the production process.

Waviness is that component of surface texture upon which roughness is superimposed. It is usually repetitive, widely spaced irregularities.

Flaws are random, relatively infrequent defects such as blow holes, scratches, or inclusions.

Lay is the predominant direction of the surface texture pattern. The combination of roughness and waviness is called profile, and it can be measured. Figure 7.10 is an instrument used to measure surface profile.

Nearly all instruments that measure surface texture operate by moving a sharp-pointed stylus across the surface. The measured profile is a representation of the profile obtained by the instrument and is distorted by the technical limitations of the instrument and by its unequal horizontal and vertical magnifications. Intentional distortion of horizontal magnification, for example, could accentuate either surface roughness or waviness.

Figure 7.10 Surface finish measuring system.

Bench Testing

Gasket applications are commonly "bench tested." This type of testing typically uses the mating flanges, along with the selected gasket, to subject the assembly to various internal pressures and temperatures. Some common bench tests are as follows.

Steam test

A steam test is used to perform short-duration thermal cyclic testing. In this test method, saturated steam and cold water are alternately circulated in the sealing passages of the gasketed joint. This quick and inexpensive test is used to "weed out" gasket designs. Auxiliary equipment used with steam testing includes

- Vacuum pumps
- Hot oil and coolant circulating systems

Figure 7.11 shows an engine being steam-tested.

Vibration table test

A vibration table is used to simulate actual unit vibration and temperature conditions. It is used with hot oil and coolant circulation system(s) and hot air. Various amplitudes and frequencies of vibration can be specified. Figure 7.12 shows a vibration table test unit.

Vibration table and environmental chamber test

A vibration table and environmental chamber system can be used to simulate actual unit conditions such as vibration and

Figure 7.11 Engine being steam tested.

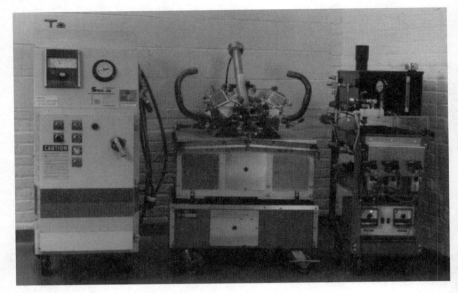

Figure 7.12 Vibration table tester.

temperature (high, low, or cyclic). This system provides the capability to set the vibration table frequency and amplitude as well as set or cycle the inside chamber temperature. In addition, hot oil or coolant, at controlled pressure(s), can be circulated through the test specimen and fixture.

A liquid nitrogen boost system can also be used to supplement the chamber's mechanical refrigeration system cool-down rate. Figure 7.13 shows this type of testing.

Oven testing

Hot oven testing is used to evaluate sealing performance of gaskets at elevated temperatures or elevated temperature cyclic conditions with actual unit components. Cyclic testing (into the oven to elevate temperature and out of the oven to ambient air or into a freezer for cool-down) provides a method to simulate material relaxation characteristics to that experienced in an actual application. The use of coolant or oil under pressure is also used in many cases to evaluate gasket sealing performance during and/or after high-temperature exposure. This type of testing is accomplished using either walk-in ovens or smaller ovens.

Figure 7.13 Vibration table and environmental chamber testing setup.

Some test devices are used in conjunction with bench testing. Some of these are as follows.

Strain gauges are used to measure and record the stretch or strain of a part or component during tensile or compressive stress applications. The installation of strain gauges onto unit components such as fastening bolts provides a very good and accurate method to monitor the initial, operating, and final loads and/or stresses experienced by the gasketed joint.

Although strain gauges are typically made from electrical resistant alloys such as constantan, they can be made of various other materials and shapes to best suit each particular application.

Strain-gauged bolts are bolts that have been equipped with strain gauges. These bolts are used to measure and record the force (static or dynamic) exerted by the bolt. Prior to use, the bolt is usually calibrated using a load frame. Strain-gauged bolts can be used to measure and record changing bolt forces experienced during unit operation, steam testing, and pulsator testing.

The limitation of strain-gauged bolts is that their diameter and length must meet the specifications for a particular application; that is, bolts requiring strain gauging are sent out for drilling and strain-gauge installation. Figures 7.14 and 7.15 show a strain-gauged bolt and a strain-measuring instrument, respectively.

A load washer is a small, strain-gauged load cell that is designed in the form of a flat washer. It is used under the head of a bolt to measure localized force (static or dynamic) exerted by the head of the bolt. Prior to use in testing, the load washer with the intended fastener is usually calibrated using a load frame. Load washers are used to measure and record changing bolt forces experienced during unit operation, steam testing, and pulsator testing.

Load washers offer some flexibility for use with bolts of different diameter and length. Limitations of the load washer are: (1) The fastener thread engagement must be long enough to accommodate the thickness of the load washer. (2) The available surface areas on the mating surface must be large enough to accommodate the load washer. (3) Use of the load washer changes the bolt grip length and therefore changes the bolt's stretch length. Figure 7.16 shows a load washer, bolt, and strain indicator.

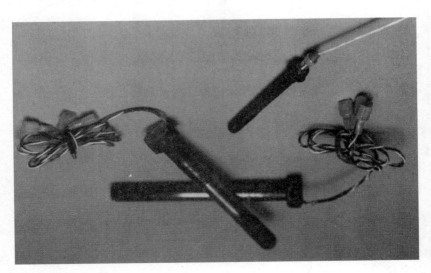

Figure 7.14 Strain-gauged bolts.

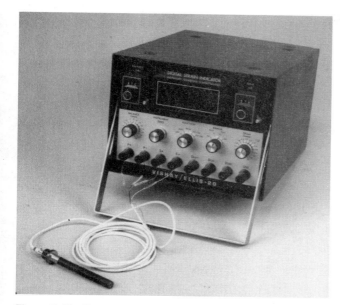

Figure 7.15 Strain-measuring instrument.

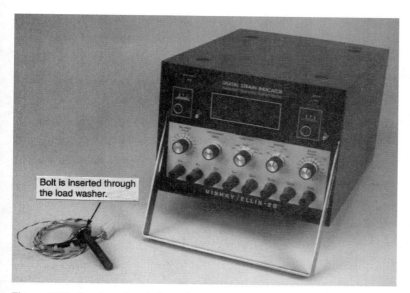

Figure 7.16 Load washer, bolt, and strain indicator.

References

ASTM Standards on Gaskets, 5th ed., ASTM 1990.

Czernik, D. E., J. C. Moerk, Jr., and F. A. Robbins: "The Relationship of a Gasket's Physical Properties to the Sealing Phenomena," SAE paper 650431, May 1965.

McDowell, D. J.: "Choose the Right Gasket material," *Assembly Engineering,* October 1981.

Oren, J. W.: "Creating Gasket Seals with Rigid Flanges," SAE paper 810362, February 1981.

Raut, H. D.: "The Effect of Flange Roughness on Leakage Rate," SAE paper 810361, presented at the SAE International Congress and Exposition, February 1981.

Zeitz, J. E.: "Signification Properties of Gasketing Materials—Pathfinders for Proper Material Selection," SAE paper presented at the SAE International Congress and Exposition, May 1968.

Chapter 8

Gasket Analysis

Failure Mode and Effects Analysis (FMEA)

As a starting point for the design of a gasket a technique, that can be utilized is failure mode and effects analysis (FMEA). FMEAs assist in problem avoidance with new or redesigned products or processes by using a structured approach to the analysis of potential failure modes. They are to be updated as design, process, or usage changes occur during the life of the product.

Design FMEAs are to be done by the design-responsible engineer during the initial stages of the design and development process, and in concurrence with manufacturing, quality, and user representatives.

Process FMEAs address process concerns and are to be done by a manufacturing engineer before the start of tooling, and in concurrence with design, quality, and user representatives.

FMEA design procedure

Purpose. The objective of this procedure is to describe when and how to create a design failure mode effects analysis.

Scope. Failure mode effects analysis pertains to the early stage of a gasket design in which potential failures are thought about. This technique should be used for each major gasket design, and it is the gasket engineer's job function to coordinate its completion.

The design FMEA supports the design process in reducing the risk of failures by

- Aiding in the objective evaluation of design requirements and design alternatives.
- Increasing the probability that potential failure modes and their effects on the part functions have been considered in the design and development process.
- Providing additional information to aid in the planning of thorough and efficient design test and development programs.
- Developing a list of potential failure modes ranked according to their effect on the "customer," thus establishing a priority system for design improvements and development testing.
- Providing future reference to aid in analyzing field concerns, evaluating design changes, and developing advanced designs.

A design FMEA should be produced for each major gasket design as part of the rubbing history for that design.

Definitions

1. Design failure mode effect analysis (FMEA)—Figure 8.1 is an example of a design FMEA. This is a comprehensive process developed to check design, manufacturing, and performance of gasket materials, with emphasis on the complete elimination of any conditions or processes that might lead to eventual product failure. It diagnoses the conditions and processes potentially conducive to failure. It presents, in writing, the prognosis for failed parts and their possible effects on the gasket operating environment. It examines the possible causes of failures before they happen and predicts possible effects that failure can have on the end product. It answers the questions "What can possibly go wrong?" and "What can we do to see that this doesn't happen in the first place?"

FAILURE MODE EFFECTS ANALYSIS, DESIGN (DESIGN FMEA)

Product Engr.: _____

Quality Assur Engr.: _____

Origination Date: _____

Gasket Maker Part No.: _____

Customer Part No.: _____

Last Revision Date: _____

Part Function	Potential Failure Mode	Potential effects of failure	SEVERITY	Potential Causes of Failure due to gasket	LIKELIHOOD	In-place and scheduled cause prevention methods	MERIT / RISK	Actions recommended to eliminate cause and to enhance prevention methods	Responsibility for action accepted by:
Seals Fluid	Fluid	on end product or user Performance loss	4	Material out of spec. thickness density	2 4	Mat'l cert. Mat'l cert.	16 32		
		Unit overheat resulting in structural damage and/or inconvenience	9	Parts manufactured out of dim. specification Port opening locations & Dims.	3 4	Initial inspect, die controlled SPC, readings each hour	36 48		
				Materials out of spec. thickness density	2 4	Mat'l cert. 1 coil per skid, FP 3016 Mat'l cert. each lot inspected, FP3016	36 72		
Seals Fluid	Fluid Swell	Extrusion and Fluid Leak	6 6	Parts manufactured out of dim. specification Port opening locations & Dims.	3 4	Initial inspect, die controlled SPC, readings each hour	81 108		
				Wrong material selection	2	Use less swell material	24		
				Wrong fluid used	4	Remove wrong fluid-replace with correct one	72		

Figure 8.1 Example of a failure mode and effects analysis (FMEA).

2. Part function—What is the product expected to do? The design is intended to meet basic functional requirements such as seal air, coolant, or combustion. Is the product intended to be a structural member, a noise isolator, a heat shield, to meter coolant, or act as a baffle? Another part function which must be included is as it affects assembly. When assembly issues are discussed, the engine manufacturer is considered the end user.

3. Failure—The result of a product's inability to perform or function properly in the application for which it was intended and designed.

4. Failure mode—The physical way in which a part fails to perform a function as intended; include items such as leaks coolant, leaks combustion, falls off engine, stops metering, or more than one gasket is installed.

5. Potential effects of failure—A functional failure of the part or aspect of the part. This would include effects of a failure on the end user and effects of a failure on the unit's builder. Include items such as overheat resulting in damage, performance loss, visual appearance, owner complaint, and stops of the production line.

6. Severity—How severe is the effect of a failure on the end user or unit builder? See the severity rating chart for appropriate values.

7. Causes of failure—What specifically could happen to or within the gasket resulting in its inability to function properly? Potential items to consider are materials, dimensions, manufacturing, handling, and functional environment. Specify cause of failure in conjunction with the significant characteristics of the gasket likely to influence the cause of failure such as material density, material thickness, embossment height, and port opening location and size.

8. Prevention method—These methods include all the quality assurance personnel, test equipment, and measuring devices installed and operating throughout the plant to ensure that the methods include periodic visual inspection, 100 percent final inspection, SPC each hour, and continuous electronic measuring with preset limits.

9. Likelihood—What are the chances that a specific "cause of failure" will result in the failure mode? Refer to the likelihood rating chart for appropriate values.

10. Merit—How good is the in-place and scheduled cause prevention method? What are the chances of this method's catching a potential failure cause before the end user receives it? See the merit rating chart for appropriate values.

11. Risk factor—The multiplicative product of the severity, likelihood, and merit values for each potential effect of a failure.

12. Actions recommended—These are planned ideas intended to answer the questions "What if the part fails, then what?" or "What methods of prevention are being investigated to improve the current method?"

13. Responsibility for action—The individual responsible for investigating the recommended actions.

Procedure

1. Create a design FMEA during the development stage of the specific major gasket design. The design FMEA is used to help determine design, manufacturing, and defect prevention methods.

2. Use the FMEA form most appropriate for the specific application. Forms are available for typical gasket applications and material types. (See applicable documents.)

3. Use the forms stored on the computer to save time and for easy updating. Save each specific FMEA on computer.

4. Complete the form header. This should be self-explanatory when reviewing the form. Note that more than one part number can be identified per FMEA, as long as the application is identical. Example: two gaskets used on the same unit, one for the left and one for the right.

5. Complete the FMEA form. Use the Definitions section of this procedure to clarify each section. Complete each column for a specific part function before progressing to the next part function.

6. Review the contents of the DFMEA with other engineering personnel. If all agree on content, then both the identity of

the gasket engineer and the DFMEA are to be part in the part folder and in the customer FMEA folder, if applicable.

FMEA form maintenance procedure

1. Use the Actions Recommended column to document necessary modifications to the product and/or prevention methods. The team should review and agree on the items listed here as well as the responsible person.

2. As part of the formal document history of a part, the FMEA should be maintained similarly to a drawing or bill of material. If a change notice is written, the FMEA form should be reviewed for necessary modifications.

3. Distribute any updates to the simultaneous engineering team. Be sure the revision date is recorded.

Severity Rating Chart

Value	Most severe result of failure
10	Human harm possible
9	Vehicle or engine damage
8	
7	Engine component damage
6	Highest level of owner dissatisfaction
5	
4	
3	
2	Lowest level of owner dissatisfaction
1	No effect on end product

Likelihood Rating Chart

Value	Predicted chance of occurrence resulting in failure
10	Each part will fail
9 8	} 1 part failure per 100
7	1 part failure per 1,000
6	
5	1 part failure per 10,000
4	
3 2	} 1 part failure per 100,000
1	1 part failure per 1,000,000 or higher

Merit Rating Chart
(Reliability of Prevention Method)

Value	Examples of prevention methods (or equivalent)
10	None
9	Periodic visual inspection
8	Periodic audits, measured
7	Preproduction measurement
6	Material samples measured prior to production
5	100% final inspection only
4	100% in-process and final inspection
3	SPC, precontrol, fixture or tool control
2	Certified and SPC and/or 100% final inspection
1	Continuous electronic monitoring with preset control limits

Finite Element Analysis (FEA)

A useful design technique that can be utilized to ensure the initial gasket design will work is finite element analysis (FEA). This technique has the following possible uses:

1. Screen the design ideas.
2. Validate the design.
3. Optimize the design.
4. Answer the technical questions without hardware available.
5. Assist manufacturing to correct the existing problem.

In general, FEA will provide a way of simulating the gasket design under working conditions and an opportunity to understand interactions with the mating mechanism. Different designs and materials can be easily changed in the model; therefore, design ideas can be screened in shorter periods of time without the real, physical components existing. FEA also provides a way of exchanging the design experience and expertise in their own discipline with the customer. Only the cooperation and combination of both can produce a qualified design for each individual application. By using FEA techniques, all design considerations and resulting difficulties quickly come to the surface. Therefore, problems in tooling or mold making would be minimized.

Some examples of instances in which FEA has been used in gasket design are:

Engine thermostat outlet gasket simulation. An FEA model of a typical water outlet gasket was built and calibrated with empirical data, and further simulations were performed on a terminal. Through FEA, one plastic material was selected and a new sealing bead was designed. This bead had a recovery capability built into the structure and was expected to perform better than conventional designs. After extensive FEA simulations, an optimized bead with better fit for narrower mating surface resulted.

Figures 8.2 and 8.3 depict the uncompressed and compressed gaskets, respectively, of this optimized bead design.

Engine valve cover design. A plastic carrier with a molded silicone rubber bead design idea was proposed. One issue was too wide of a span of the cap screws on the thin sheet metal valve cover. In the middle span the gap was measured under clamped conditions and found to be excessive. Therefore, in order to seal this joint, a superhigh bead had to be proposed. Because the bead was unusually high, it tilted instead of flattening out under clamping load. This resulted in two questions: (1) If it is tilted, what would the stress pattern look like? (2) Would it generate any problems that would have to be considered? Through FEA simulation these questions were answered. Figures 8.4 and 8.5 are the FEAs associated with the work.

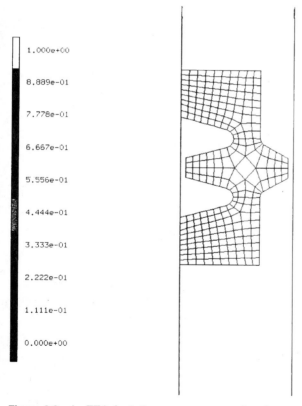

Figure 8.2 An FEA depicting an uncompressed gasket.

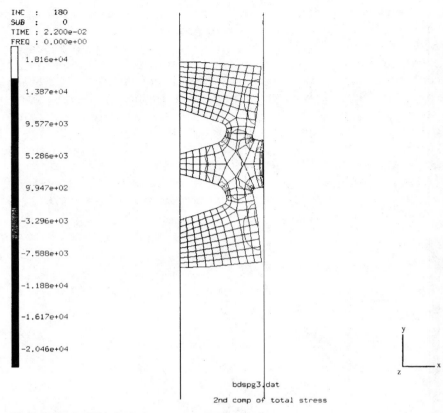

Figure 8.3 An FEA depicting a compressed gasket.

Computers and gaskets

Various gasket material manufacturers have developed computer programs for gasket selection purposes. These programs are essentially databases where gasket experience and material characteristics are stored and used in "if–then" regulations. Most of the systems perform preliminary calculations concerning the clamping pressure using the environmental conditions of member and size of m fasteners, torque on them, and gasket clamping area. Other factors that are usually considered are:

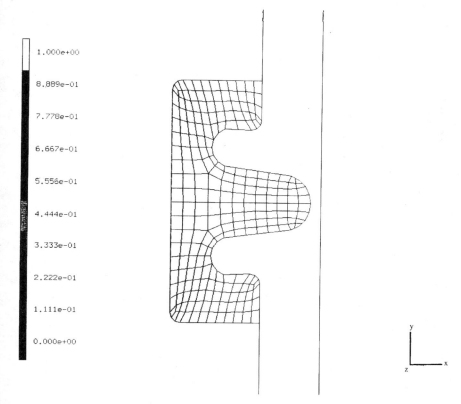

Figure 8.4 An FEA of the uncompressed molded bead.

Application operating temperature
Application operating pressure
Application medium to be sealed
Flange flatness and roughness
Sealing performance required

As might be expected, these programs consider the particular gasket material that the manufacturer produces. The gasket designer should contact his or her gasket suppliers in regard to their computer gasket programs.

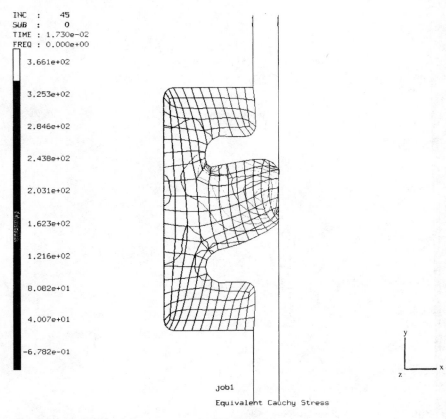

Figure 8.5 An FEA of the compressed molded bead.

References

Bathe, K-J., *Finite Element Procedures in Engineering Analysis,* Prentice-Hall, 1982.
Huston, R., L. and C., E. Passehello, *Finite Element Methods, an Introduction,* Marcel Dekker, Inc., 1984.
"Potential Failure Mode and Effects Analysis (FMEA)," Ford Motor Co., September 1988.
"Procedures for Preforming a Failure Mode, Effects and Criticality Analysis," MIL-STD-1629A, November 24, 1980.

Chapter 9

Gasket Leakage and Chemical Gaskets

Leakage—Detection and Rating

Introduction

Elimination or reduction of leakage is the main reason for the use of gaskets. Fluid dynamics has analytically shown that leakage in most gasketed joints can be mathematically defined as follows. See Fig. 9.1.

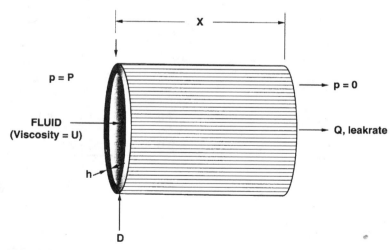

Figure 9.1 Fluid dynamics leakage rate.

$$\text{Leak rate} = Q = f\frac{D \times h^3 \times P}{Ux}$$

This is associated with a circular leakage path

where Q = leakage rate in volume per unit time
 D = diameter of gasket
 h = leakage path between gasket and leakage diameter
 P = pressure differential between system pressure and environment
 U = viscosity of sealed media
 x = length of leakage path

Rating systems

A number of techniques are used for leakage detection. The techniques range from the low-tech visual means to the high-tech helium leakage detection instruments. A rating criterion associated with visual observations is shown in Table 9.1.

Another rating standard that has been used in the transportation industry is depicted in Fig. 9.2.

The following chart shows the amount of fluid loss that is associated with various liquid leakage rates.

	Fluid loss, gal		
Leakage rate	Loss per day	Loss per week	Loss per month
1 drop every 10 s	0.1	0.7	3.0
1 drop every 5 s	0.2	1.4	6.0
1 drop per s	1.00	7.0	30.0
3 drops per s	3.00	21.0	90.0
Drips breaking into a stream	20.0	140.0	600.0

Leak-detection means

The most basic leakage detection is visual observation. In order to improve this observation, powders such as spray foot powders or kitchen cleansers mixed with water have been utilized.

Gasket Leakage and Chemical Gaskets

TABLE 9.1 Leak Rating Designations and Definitions

Leakage class	Class definition
0	No physical evidence of leakage
1	Dampness can be seen or felt, but no droplets formed
2	Localized droplets formed
3	Droplet bead formed the entire length of gasket's edge that can be seen
4	Droplets form and flow from gasket's edge to generate a drip that falls at a rate of 1 drop in 5 or more minutes
5	Drops fall at a rate of 1 drop in 5 or fewer minutes
6	Steady stream of fluid

LEAKAGE CLASS	CLASS DEFINITION	VISUAL APPEARANCE
0	NO PHYSICAL EVIDENCE OF FLUID	
1	A NON RECURRING FLUID FILM AND/OR CAPILLARY ACTION	
2	A RECURRING FLUID THAT DOES NOT RESULT IN THE FORMATION OF A DROPLET, AND WHICH MAY ALSO BE CAUSED BY SEEPAGE.	DAMPNESS BUT NO DROP FORMATION 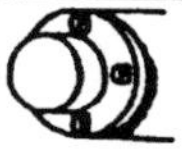
3	A RECURRING FLUID THAT RESULTS IN THE FORMATION OF A NON FALLING DROPLET, OR WHICH DRIPS AT A RATE OF ONE DROPLET IN FIVE OR MORE MINUTES.	DAMPNESS DEVELOPING A DROP 
4	A RECURRING FLUID WHERE A DROPLET FORMS AND FALLS IN LESS THAN FIVE MINUTES.	
5	A RECURRING FLUID WHERE THE FREQUENCY OF DROPLETS MAKES A MEASURABLE STREAM.	

Figure 9.2 Leakage rating system.

More exacting means, such as powders that glow under ultraviolet light, have been used in recent times.

Tracking small, elusive leaks by using conventional methods is a time-consuming, labor-intensive chore that often proves futile despite the enormous effort. Tricky air currents, irregular component shapes, design-limited visibility, and the drawbacks of conventional detectors can make it difficult to identify the type of leak, much less to pinpoint the exact source.

Other leak-detection techniques have disadvantages. Visual inspection enables the operator to find only the most obvious leaks and cannot be used for small leaks that evaporate (or burn off) on contact. Compressed air, used only in water-cooling systems, can actually cause additional leaks, can enlarge existing ones, can damage hoses and radiators, and can even blow gaskets if not used properly.

Soap solutions are not the answer, either. This messy, time-consuming technique will not find large or very small leaks and is useless in detecting leakage of oil, automatic transmission fluid (ATF), fuel, and hydraulic fluid.

Although spraying talcum powder on a suspect area can occasionally be used to locate a leak, using it in drafty areas or while the unit is running will make a real mess. Furthermore, great care must be taken to keep this abrasive powder from contaminating the sealed fluid or the leak may become the least of your problems.

Halide torches and electronic detectors (also called ionizers or sniffers) are used to detect leakage. The torches present a burn and fire hazard, produce strong odor, and emit gases harmful to human tissue, especially the respiratory system. These must be considered before the torch is used. In addition, electronic testers are delicate instruments that can be rendered inaccurate or inoperative by rough handling, heat, humidity, or capacitance.

There also are (1) rate of pressure change and (2) mass-flow leak-detection devices. One measures the rate of pressure change; the other directly measures leakage mass flow. In the rate-of-change method, the test part is pressurized and then isolated from the pressure source. Any change in the part's pressure with time will permit calculation of a leakage rate. (Uncontrollable variables such as ambient temperature changes, drafts, test-part deformation, or seal creep can cause problems.) In the mass-flow method, the test part is pressurized and any leakage is compensated naturally by air flowing into the test part from the source; the amount of air that flows in to replace leakage flow is measured directly in standard volumes per minute.

This used to be viewed as slower and less reliable than pressure-change detection. Recently improved mass-flow sensor technology, coupled with the use of microprocessor-based electronics and control reservoirs, has dramatically raised the performance of these systems. The reservoir allows the test to be isolated from the pressure source by serving as an alternative source that is more stable than earlier conventional pressure regulators.

Pressure-change systems require measurement of test part pressure at two different times to calculate the change rate. An error in either measurement results in an equivalent or larger error in the required difference calculations. Owing to other variables, the probability of error increases with the measurement, which is generally more accurate and completed in much less time (typically <1 s), minimizing the impact of uncontrolled variables.

The difference in performance between these two methods is especially important when there is a need for true volumetric leak data on both accepted and rejected parts as input to SPC programs, and the continuing drive toward advanced capability systems.

Improved performance enables the mass-flow leak detector to be used as a true measurement instrument capable of being evaluated as a precision gauge. Repeatability and reproducibility (R&R) studies in typical applications of computerized mass-flow leak-detector systems, including the test fixture and seals, yield consistent R&R percentages of <10 percent for most applications.

Pinpointing the source of a leak can be tricky. One should not automatically assume that the leak is originating at the unit even if that's where one has been seeing most of the leakage accumulation. For instance, suppose the unit flange is porous in the area just above the suspected leak. Flanges with porosity problems don't spew out large volumes of fluid. Rather, they weep or "sweat" fluids in amounts so minute they're virtually undetectable with the naked eye. Gravity then does its bit by pulling the fluid down onto the unit, where it eventually runs over the sides. What one ends up with is a classic case of misdiagnosis.

In the old days, tracing a leak source was pretty much a hit-or-miss proposition. You'd coat the area with powder and start the unit to get everything pressurized. If things went your way, the leak would show up as a dark spot in the white powder. Today, spotting the source of a leak is more exact and less cumbersome, thanks to the black light leak-detection system on the market. The idea behind these systems is very basic. A small amount of fluorescent dye is added to the sealed media. The unit is then started to build pressure and allow the dye to circulate. After a few minutes, one simply aims the light on the suspect area and the leak shows up as a bright glow.

The ultraviolet (UV) leak-detection system is effective. It is used and approved for service application by major vehicle manufacturers worldwide. With a high-density, long-wave UV lamp and fluorescent additives, leaks as small as 0.25 oz/year (or even less) can be quickly and easily found in any enclosed circulatory system. A bright fluorescent glow instantly reveals the sources of all leaks in the units.

Although ultraviolet or black light leak detection is not exactly a new concept, compared to other methods of oil, coolant, power steering, transmission fluid, and A/C leak detection it certainly qualifies as state-of-the-art (with the possible exception of electronic halogen A/C leak detectors), especially if you're used to using foot sprays, soap and water, newspapers, and pressure testers.

With the black light system, you have the capabilities of developing a systematic and patterned approach to finding virtually all types of leaks, since all types of leaks are checked in the same manner. Once you have the basic light unit, all you need do is purchase the specific fluorescent dye additives that are compatible with the fluid system you're testing.

Once you've determined the nature of the leak you are diagnosing, place the dye additive into the system you are checking. After the dye has had a chance to mix and circulate throughout the entire system, the light is used to scan all potential leak areas including hoses, fittings, gaskets, etc. If a leak is present, the black light's ultraviolet rays will make the dye glow a bright yellow. To pinpoint the leak's origin, all you do is trace the "glow" back to the source.

Gasket Fabrication

Nonmetallic gaskets are converted from either sheets or coils. The typical tools and/or equipment used to produce gaskets are as follows.

Steel rule dies. These dies are most commonly used for moderate-volume parts. Tooling is economical and can be used in platen presses, roll presses, mechanical presses, or rotary die machines. The dies are generally used when cutting nonmetallic gasket materials where gasket dimension tolerances are greater than ± 0.015 in. Rule dies can be used for low- or high-volume parts when cutting nonmetallic materials. These dies also can be used on metals of light gauge. For metallic gaskets, this type of tooling is generally used for prototype or low-volume parts.

The dies are generally made as follows:

- Standard jig cut steel rule die: Hand layout on the base board or onto a mylar which is then adhered to the board. The holes are drilled and contour sawed on a jigsaw. The rule is bent and fitted into the slots.

- Standard jig cut and precision bored rule die: The operations are the same as above except the holes are drilled in a precision machine such as a mill, jig borer, or N/C machine.

- Most precise rule die: The part is programmed into a laser cutting machine. The machine cuts out the hole and the contour to a precise dimension. The rule is bent by hand and pressed into the base board.

Sharpened hole punches known as "tubes" are incorporated along with rubber stripper pads to complete the tool. Depending upon the dimensional aspects of the part, different procedures can be used in the construction of the cutting tool. These various methods greatly dictate the accuracy of the cut part.

Figure 9.3 depicts the types of rules used.

Table 9.2 shows the dimensional capabilities for steel rule dies.

Steel male and female blanking dies. These dies are either progressive dies or compound blank dies consisting of male and

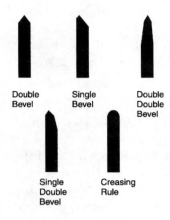

Figure 9.3 Various types of rule dies.

female die components. The tooling is considerably more expensive than rule dies but is more accurate and can hold tolerances of $\pm$ 0.002 in. These dies are used for high-volume parts, tight tolerance parts, and for thicker metals—0.010 in and thicker.

The compound die produces the most accurate part and is generally less expensive than the progressive die. Sometimes, because of narrow walls or processing problems, this type of tool cannot be used and a progressive die is used.

Electrical discharge machining (EDM). This is done with an electrical discharge traveling wire machine. Only ferrous metals can be cut using this process. It is a highly accurate method of cutting parts. The parts are stacked generally 1 to 2 in and machined together. This method is good for prototype or very low volume parts but is rather slow and expensive.

Laser cutting. This is an inexpensive method to machine parts. The cutting speeds are extremely fast, approximately 8 yd/min, and one can cut materials that are stacked, thereby producing higher production rates. Here again nesting of parts for best material can be done.

Water jet cutting. This cutting method is relatively new to the gasket industry. It is extremely fast and similar to laser cutting in that no tooling is required. Cutting is determined by computer programming.

TABLE 9.2 Dimensional Capabilities for Steel Rule Dies.

Tooling Feature Linear Dimensions	Tooling Type		
	A	B&C	D
<1" to 23.999" 24.000" to 49.999"	± .015 ± .031	± .010 *1 ± .020 *1	± .007 ± .015
Punched Holes *2 .999" & below 1" & above	+ .015 - .005 ± .020	+ .015 - .005 ± .020	+ .015 - .005 ± .020
Hole Positional Tolerance based upon maximum Material conditions (M.M.C.)	.030	.015	.015
Radii .062" to .249" .250" to .499" .500" to -	± .030 ± .020 ± .015	± .030 ± .020 ± .015	± .030 ± .020 ± .015
Angular	± 1°	± 1°	± 5°

< Designates less than

Note" *1 - For rectangular coordinate spacing of holes only. Other linear dimensions to have same tolerance as A-1.

*2 - For material thickness of .093" and below. Values do not include taper caused by breakout or from bevel of punch tube.

Information from "Technical Handbook" of Gasket Fabricators Association

Gasket dimensions

The dimensions and tolerances associated with gaskets are determined by the cutting tools and the gasket materials. It is important that the critical dimensions of the gasket be clearly identified. Generally dowel and bolt hole location are easier to control than contour.

Contours and part holes are dependent on hand jigging and hand rule bending. The gasket thickness also has an effect on dimensional tolerances. Gasket manufacturers should be consulted in regard to the plan dimensions and tolerances.

Dimensional tolerances that can be held for various gaskets are dependent on the material from which the gasket is made. Measurement of a gasket's dimensions is not conducted on the gasket itself if the gasket is unstable or flimsy. In these cases, the measurements are made using stable materials blanked with the gasket's tooling.

In general, thicker gaskets have greater thickness tolerances than thinner gaskets. Table 9.3 depicts the typical thickness tolerances associated with various materials.

The following are measuring devices commonly used in the gasket industry:

- Calipers
- Coordinate measuring machines
- Durometer gauge—instrument to check hardness of rubber and rubberlike material
- Gauge pins—straight, unflanged pins with specific diameters and extremely close tolerances
- Height gauge—check basic X and Y dimensions
- Light-section microscope
- Metal hardness tester—device to determine hardness of steel being fabricated
- Micrometers
- Optical comparators
- Radius gauges—precision ground metal strips with accurate radius machines on each end
- Scales—6, 12, and 18 in
- Shadow graph machines
- Templates—soft (thin plastic or mylar) and hard ($\frac{1}{8}$- to $\frac{1}{2}$-in-thick plastic or mylar with steel pins)
- Tolerance gauge—tool for visual pass-fail dimensional inspection.

Gaskets are identified using a variety of means. The following are some of the more common ones:

TABLE 9.3 Typical Thickness Tolerances Associated with Nonmetallic Gasketing Materials.

Typical Thickness Tolerances (Inches)*
Non-Metallic Gasketing Material

THICKNESS

Product/Process	.016"	.031"	.047"	.063"	.094"	.125"
Beater Addition Fiber	+.002 / -.003	±.003	±.004	+.004 / -.006	+.010 / -.006	+.011 / -.009
Compressed Sheet	+.005 / -.002	±.005	±.005	±.008	±.008	±.010
Cork Composition	±.010	±.010	±.010	±.010	±.010	±.013
Cork - Rubber Rolls	-	±.005	±.005	±.005	±.010	±.010
Sheets	-	±.010	±.010	±.015	±.015	±.015
Expanded Graphite	±.002	±.002	±.002	±.002	±.009	±.013
Rubber	±.010	±.012	±.012	±.016	±.016	±.020
Rubber Coated Fabric	±.003	±.003	±.005	±.006	±.009	±.013
Sponge	-	-	-	+.032 / -.016	+.032 / -.016	+.032

* Derived from published literature. ASTM F 104/SAE J90 Standard Classification System, and RMA Sheet Rubber Handbook. Acutal Tolerances may vary by manufacturer. Some manufacturers offer tighter tolerances at premium prices. Verify with supplier CPK level attainable with supplier's process.

Information from "Technical Handbook of Gasket Fabricators Association"

1. Rubber stamp with ink—part number and/or supplier log
2. Metal stamp—no ink—indent material with part number and/or supplier logo
3. Screen print—part number and/or supplier logo
4. Tie in bundles and tag
5. Package specific quantities in printed envelopes
6. Shrink pack specific quantities—label
7. Color code with rubber stamp or screen printing
8. Notch edge of gasket according to prearranged code

Chemical Gaskets

Chemical gaskets are newcomers to gasketing but have achieved considerable acceptance and adoption in the field. Chemical gaskets are available in liquids of varying viscosity, pastes, puttylike mastics, and tapes dispensed manually or automatically from bottles, tubes, and cartridges. Chemical gaskets have achieved considerable acceptance and adoption in industrial gasketing. The two materials that have emerged in this regard are room temperature vulcanizing (RTV) silicones and anaerobics.

Part A

The RTV silicones cure by absorbing moisture in the air and giving off acetic acid; the newer types give off amines. RTV silicones having a low content of silicone volatiles are available. These volatiles may foul sensors used in engine emission systems. The anaerobics cure in the absence of air when in contact with an active metal, i.e., when air is excluded as in a clamped gasketed joint.

Both classes of materials contain 100 percent solids, ensuring "no voids" or shrinkage occurs during drying or curing. Since the mating flanges are assembled when the material is uncured or wet, shrinkage during drying and curing could result in leak paths. Other types of sealant materials are available but they are restricted to limited applications and therefore are not covered in this handbook. In some industries, chemical gaskets are called formed-in-place gaskets.

The theory of sealing of chemical gaskets differs substantially from that of typical mechanical gasketing. In the latter case, the material is compressed and exerts a sealing force on the flange proportional to its stress-retention and recovery characteristics. Owing to the support of the load, the gasket tends to follow the motion of the mating flanges whether they be thermal or mechanically induced. In some applications bolt stretch is a factor in maintaining joint tightness.

In chemical gaskets, metal-to-metal or flange-to-flange contact occurs and the gasket does not support load or contribute to tolerance stack-up. Chemical gaskets seal by filling the gaps between the flanges and adhering to the flanges.

Since the material is not under load, it has no desire to follow the flanges as they try to separate unless it is bonded to the flanges, and the adhesion is strong enough to prevent motion, or the gasket material has enough extensibility to follow the motion without breaking the bond. Silicone RTVs are rubber and do have high extensibility. However, the thickness between the flanges is very low. With a percent elongation at breaking point of 400 percent, the practical amount of extensibility is less than four times the thickness. Silicone RTVs are also low-strength materials with limited ability to prevent motion. The anaerobics have very low extensibility, but much higher strength than silicones and thus a greater ability to prevent motion between the flanges.

The high extensibility of silicone RTVs allows them to follow flange motion. To ensure adequate silicone gasket thickness to follow motion, it is a common practice to include gasket stops designed into one of the mating flanges. In some cases, the flange is grooved to permit a thick silicone RTV bead. Applications where the flanges are subjected to cyclic separating forces whether dynamic or thermal should be thoroughly evaluated before chemical systems are adapted.

In the case of anaerobics, they seal by completely filling the gap and adhesively stopping the flange motion by bonding the flanges together. Anaerobics have little extensibility but do possess very high shear and tensile strength. Bond strength is a controlled variable. Ease of disassembly and flange clean-up are a function of bond strength. They are used for heavy cast flanges and tight-fitting cures, such as oil seal cures.

When metal-to-metal contact occurs with chemical gaskets, torque loss is normally negligible. In addition, flange motion is generally less than that in the case of a mechanical gasketed joint, since the fastener torque is retained. Also, flange distortion is minimized, since the flanges are assembled while the chemicals are uncured and metal-to-metal contact occurs. Chemical gaskets may not contribute to tolerance stack-up.

Part B

With chemical gaskets, one must be concerned not only with the cure properties of the material but with the uncured properties as well. We have already mentioned the importance of 100 percent solids content.

One concern is the viscosity of the chemical gaskets. The chemicals must fill the gap as well as the unevenness of the mating flanges and remain in place until cured. Therefore, the viscosity must be high enough to prevent flowing out. In the case of anaerobics, for example, intimate contact with the mating flanges is required to ensure exclusion of air, curing, and adhesion. Therefore, the viscosity of the compounds will be dictated by the gap thickness that needs to be filled.

To completely fill the gap, an excess amount of material is required. This, however, can result in the material's flowing into the cavity to be sealed, and mixing with the sealed media. This possibility must be considered when specifying chemical gaskets. There have been cases where the excess material has caused problems by blocking passageways in the assembly.

The most popular silicone RTVs on the market cure by absorbing moisture from the air; therefore, their curing characteristics and rate of cure are a function of relative humidity. Figure 9.4 is a graph depicting the drying time (rate of cure) versus percent relative humidity for an RTV compound. Note that 20 percent and higher relative humidity is required for rapid cures.

Figure 9.5 depicts the set time versus the gap size for an anaerobic sealant. Note the difference in both set time and gap

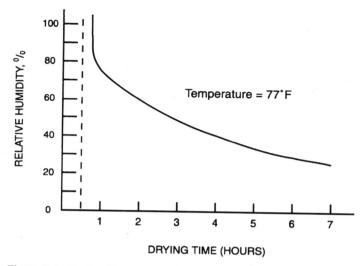

Figure 9.4 Drying time versus relative humidity for an RTV silicone.

size associated with the cleanliness of the steel and the use of primers.

There is a limit to the size gap that can be filled and still result in the cure of the anaerobic. The activity level of the flange metal also affects the rate of cure. In addition, depending upon the internal pressure and the time permitted for curing, there are limitations on the size of the gap permitted for both silicone RTVs and anaerobics.

Concerning storage stability, the RTVs must be kept tightly sealed until just prior to use. Anaerobics conversely do require exposure to oxygen in order to prevent cure; therefore, the storage containers must take these into account. Most chemical gaskets have about a 1-year shelf life.

The following are some concerns associated with uncured and cured properties of chemical gaskets.

Uncured properties

1. Percent solids—Any solvents or water in a chemical gasket will evaporate, causing shrinkage and possibly leakage. Silicone RTVs and anaerobics are essentially 100 percent reactive material, and no shrinkage occurs because of evaporation.

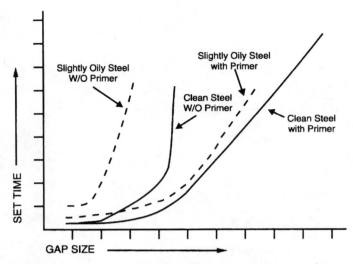

Figure 9.5 Set time versus gap size for an anaerobic material.

The shrinkage due to the chemical reactions leading to the formation of a solid gasket material is less than 0.1 percent.

2. Storage stability—Silicone RTVs cure by reacting with the moisture in the air; therefore, they must be kept tightly sealed until just prior to use. Anaerobics, conversely, require exposure to the oxygen found in air to prevent gelation. Low-density polyethylene is highly permeable to oxygen and is the preferred material for packaging anaerobics. Properly packaged and stored quantities of silicone RTV and anaerobics have shelf lives over 1 year.

Die-cut or preformed gaskets have a much longer shelf life. In fact, stored gaskets cut or preformed prior to the invention of silicone RTVs and anaerobics are still functional. This is an important consideration in OEM service and aftermarket planning.

3. Viscosity—Since the gasket must fill the gap and the unevenness of the mating surfaces, its viscosity or consistency must be high enough to prevent flowing out. In the case of anaerobics, intimate contact with the mating surfaces is required for exclusion of air and subsequent curing. Gap thick-

ness will dictate what type and what viscosity of anaerobic can be used.

4. Compatibility with the fluid being sealed—In order to completely fill the gap, it is not only desirable but necessary to use an excess of material. Ideally, the excess bead of material will either cure in place or be soluble or dispersible in the fluid being sealed. The effects of uncured or cured chemical gasket material in the system should be evaluated. In most cases, the bead of excess material will cure in place, provided the amount used is controlled. Uncontrolled applications resulting in the material blocking liquid passageways have resulted in failures. This is a problem that does not exist with precut or preformed gaskets but must be considered in the case of chemical gaskets, especially in the repair situation.

5. Cleanup—Chemical gaskets are wet. Drips and spills can and do occur. In the case of silicone RTVs, they can be wiped up with a rag or the material can be allowed to cure and then removed. In the case of anaerobics, the drip and spills will not cure, because air is not excluded. They must be wiped up with a rag. Anaerobics will also attack and lift some coatings.

Most die-cut or preformed gaskets can be dropped on the floor, picked up, and used. They are neater.

Curing process

1. Silicone RTVs—These require humidity to cure. The moisture in the air reacts with the polymer, releasing acetic acid, opening reactive sites, and allowing the liquid polymer to cure to a solid silicone rubber. Cure time is related to relative humidity, and the thickness to be cured. Cure speed is very slow at relative humidities below 20 percent. Very dry conditions are not so uncommon that this should be ignored.

2. Anaerobics—These cure by the exclusion of air and the accelerating or catalytic effect of an active metal surface. Cure speed depends on the gap thickness, the active metal surface, and the cleanliness or prior treatment of the metal surfaces. Anaerobics will cure on lightly oiled surfaces, phosphated surfaces, and other surface treatments. However, cure time will be different for each surface.

For both materials, the important factor is not time to complete cure but time required to cure enough for sealing. In some cases, only gelation is required. In other cases, complete cure is required for sealing the internal pressures. Die-cut or preformed gaskets do not have a cure time limitation. However, some applications do require "running in" and retorquing.

Cured properties

1. Chemical resistance—As with die-cut or preformed gaskets, chemical resistance is of utmost importance. Silicone RTVs, while not having outstanding chemical resistance, are resistant enough for use against most oils and coolants. Anaerobics have outstanding resistance to oils, coolants, and many solvents.

2. Adhesion to the flanges—It seems obvious that if the gap between the flanges is completely filled, and the material used to fill the gap is firmly bonded to the flanges, a seal is formed. This is a key mechanism of chemical gasket technology. The adhesive strength and the chemical resistance of the bond are key properties. Adhesion after use must be considered in regard to cost of repair. Many precut gaskets have antistick coatings applied for ease of removal.

3. Strength—Of course, the higher the strength of the cured material, the higher the pressures it can seal. Silicone RTVs are weak materials in comparison with anaerobics or die-cut gaskets and should not be considered for high-pressure applications.

4. Compressibility, extensibility, and flexibility—These properties are a measure of the ability of the chemical gasket to follow and survive the movement of the vibrating flanges. Silicone RTVs have a high degree of compressibility, extensibility, and flexibility. Anaerobics have little compressibility and extensibility. However, they can be formulated to have adequate flexibility for most applications.

5. Heat resistance—The operating temperature range for RTV gaskets is −100 to 600°F. Anaerobics are suitable from −60 to 400°F. These ranges are for continuous duty.

6. Torque retention—The metal-to-metal configuration of chemical gaskets results in virtually zero torque loss in most

cases. Precut gaskets have wide ranges of torque retention properties.

Part C

Another consideration in sealing with chemical gaskets concerns the dispensing of the material and the cleanliness requirement of the mating flanges. The chemical gasket must be firmly bonded to the flanges for a seal to be achieved. The adhesive strength of the bond is a key property. In addition, compressibility, extensibility, and flexibility of the products are also important. These properties are measured by the ability of the gasket to follow the movement of the mating flanges. Silicone RTVs have a high degree of compressibility and extensibility, while anaerobics have little extensibility and compressibility. Enough material must be applied to the mating surfaces to fill the joint gap. Squeezing it out of a tube with a nozzle on the end leaves room for operator errors. To remove the operator error, anaerobics can be accurately applied by silk screening or stenciling. RTVs, because of their sensitivity to moisture, can be applied with a pantograph or tracing type machine.

After application of the material, the flanges must be carefully assembled to avoid wiping the bead off. These materials harden after assembly. If a leak develops, the unit must be disassembled and the sealing operation repeated. In addition, there is a difference in the ease of application on the original build where automated equipment is permissible versus the repair case where hand application is the only possible means of applying the chemical.

The cleanliness of the flanges is not usually a problem in original equipment (OE) builds but can be a major consideration for service. In some cases, the original build gaskets may be a chemical while the service gaskets may be mechanical primarily because of the service-related requirement. Service requirements such as inventory, installation, time, dispensing difficulty, and shelf life are all considerations in determining if the chemical gasket should be changed to a mechanical gasket for service.

RTVs generally can accommodate gaps up to 0.1 in. Anaerobic sealants by comparison generally can accommodate gaps up to 0.01 in. The use of primers allows larger gaps to be sealed in some cases.

Selection of a mechanical or chemical gasket for a given application depends upon a number of factors. When selecting, it is perhaps easiest to look at those applications where chemical gasketing is not suggested:

- The temperatures are too high or too low for the chemicals.
- The chemicals and medium being sealed are not compatible.
  Medium attacks chemical
  Chemical contaminates medium
- The thermal and/or mechanical motion is too high for the chemicals.
- The gasketed joint may have to be pressurized before the chemical can cure adequately.
- The gasket must act as a shim, metering, or blocking device.
- Metal-to-metal contact is not permitted for noise or other reasons.
- The visual appearance of the joint precludes use of chemicals.

The following are some chemical selection guides.

Guidelines for anaerobic gaskets

- Dowels and bolt holes to be chamfered to eliminate raised metal and shimming.
- Apply material inside or around dowels and bolt holes to eliminate leak paths.
- Assemble components 1 h after application of material to minimize contamination.
- Torque fasteners to specifications within 3 min of application.
- Allow 30 min cure time before application of pressure.

Guidelines for RTV silicone gaskets

- Ensure flange cleanliness.
- Avoid exposure of silicone to contamination.
- Dowels and bolt holes to be chamfered to eliminate raised metal and shimming.

- Apply material inside or around dowels and bolt holes to eliminate leak paths.
- Cure in humid environment.
- Assemble within 3 min (50 percent relative humidity at 70°F).
- Allow 30 min cure time before application of pressure.

The operating pressures that chemical gaskets can accommodate are dependent on several factors such as cure time, joint width, gap, and clamp load. In general, RTVs are not normally recommended for high-pressure applications. Owing to the small gap associated with sealing with anaerobics, and their high strength, higher pressures can generally be accommodated by this class of chemical gaskets.

Advantages of chemical gaskets

1. Eliminate the need for specific cut gasket inventories.
2. High microsealing capabilities.
3. Eliminate compression set and subsequent torque loss.
4. Able to be applied to parts in horizontal, vertical, and overhead positions.
5. Easily applied on automatic, semiautomatic, or manual dispensing equipment.
6. Do not require cutting tools.

Disadvantages of chemical gaskets

1. Some of them create by-products which can cause corrosion of the mating flanges and/or the assembly.
2. In some cases, the working time is dependent on temperature and humidity.
3. Has a limitation of gap-filling ability.
4. Have limited flange motion following ability.
5. Some chemical gaskets can cause some skin irritations.

Chemical gaskets definitely have a place in industrial gasketing applications. Their sealing ability, however, varies greatly with the chemical compound itself. It is therefore recommended that the material manufacturers be consulted when specifying a chemical gasket for a specific application. in addition, there is a "Design Guide for Formed-in-Place Gaskets." It is SAE Publication J-1497 published in May 1988. It contains information on the types of sealants available, cure systems, cured and uncured properties of the chemical gaskets, as well as application techniques for the initial seal and for surface resealing.

References

"ASTM Standards on Gaskets," ASTM 1990, fifth ed.
"Chemicals: Wonder Drugs or Cure-Alls?" *Import Car,* February 1983.
Czernik, D. E.: "Recent Developments and New Approaches in Mechanical and Chemical Gasketing," SAE paper 810367, February 1981.
Czernik, D. E., J. C. Moerk, Jr., and F. A. Robbins: "The Relationship of a Gasket's Physical Properties to the Sealing Phenomena," presented at the SAE International Congress & Exposition, May 1965, paper 650431.
Freudenberger, R.: "Gaskets or Glue," *Motor Service,* April 1982.
McDowell, D. J.: "Choose the Right Gasket Material," *Assembly Engineering,* October 1978.
Nolan, D.: "Chemical Gasketing for Automotive Powertrains," *Automotive Engineering,* August 1993.
Oren, J. W.: "Creating Gasket Seals with Rigid Flanges," presented at the SAE International Congress & Exposition, February 1981, paper 810362.
Raut, H. D.: "The Effect of Flange Roughness on Leakage Rate," presented at the SAE International Congress & Exposition, February 1981, paper 810361.
SAE J1497 "Design Guide for Formed-in-Place Gaskets," 1988.
Zeitz, J. E.: "Signification Properties of Gasketing Materials—Pathfinders for Proper Material Selection," Presented at the SAE International Congress & Exposition, May 1968, paper 680500.

Chapter 10

Engine Gaskets

Internal Combustion Engine Gaskets

Gaskets for the internal combustion engine appear simple but in reality are highly engineered products. A gasket design is selected for use only after extensive functional testing. Engine dynamometer testing, usually at overrated power outputs and thermal shock conditions, along with long-term field testing, are almost always conducted before the gaskets are adopted for use. Prior to the engine manufacturer's expending the costs involved in this testing, the gasket manufacturer must show that engineering analysis and testing supports the expenditure.

In theory, if the flanges were perfectly smooth and parallel, and infinitely rigid, one could bolt them together and seal without a gasket. But, in practice, flanges have rough surface finishes and limited rigidity. In addition, flange loading is often nonuniform across the flange surface. Gaskets, therefore, are introduced in the joint to maintain sealing by:

- Compensating for the nonuniform flange loading and flange distortion.
- Conforming to flange surface irregularities. Engine head gasketing is particularly indicative of nonuniform flange loading.

Table 10.1 depicts the environmental conditions associated with the sealing of internal combustion engines.

TABLE 10.1 Environmental Conditions of Internal Combustion Engine Sealing

System	Media	Environment[1]		Motion
		Temperature °F	Pressure[2] psi	
Combustion System	Combustion Gas	2000	1000 to 3000	Dynamic deflections, Thermal differential
Coolant System	Water/Glycol or Treated Water	-40/280	20 30 Perf., 50 Marine	Thermal differential combinations
Lubrication System				
1. Oil Sump	Engine Oil	-40 300	Vacuum to 5 psi	Thermal differential
2. Oil Inlet - Suction	Engine Oil	-40 300	Vacuum	Joint motion may occur during vehicle operation
3. High Pressure Oil	Engine Oil	-40 300	80	Thermal differential
Intake Air System				
1. Compression Ignition	Air			
Naturally aspirated		Ambient	Vacuum	Thermal differential
Turbocharged		5/320	15 to 30	
2. Spark Ignition				
Carburetor	Air/Fuel	-40/130	Vacuum	Thermal differential
Fuel Injection			Vacuum	
Tubocharged			20 to 40	
Exhaust System				
1. Compression Ignition				
Naturally aspirated	Combustion Exhaust	1300	5	Thermal differential
Turbocharged		1300	10	
2. Spark Ignition				
Carburetor	Combustion Exhaust	1800	5	Thermal differential
Fuel Injection		1800	5	
Turbocharged		1800	5	

[1] Includes High Performance Engines
[2] Pressure surges and cycle to cycle fluctuations of varying magnitudes will occur during operation and especially at cold starting JAS0060

The cylinder head gasket is the most critical sealing application on any engine. Typically, it must simultaneously seal: (1) high combustion pressures and temperatures; (2) water and antifreeze with its high wicking and wetting characteristics; and (3) lubricating oil with its associated detergents, additives, and variable viscosities either "built in" or changed with the season. In addition, the head is a structural component of the engine; i.e., the combustion chamber is formed by the head, block, piston, piston ring, and gasket. The gasket shares the same strength requirements as the other combustion chamber components.

The head gasket is used many times to either meter or block coolant flow for proper cooling of the engine. It also seals the block-liner intersection in wet liner engines. Its compressed thickness affects the compression ratio of the engine, and the importance of compression ratio control and emissions, especially in diesel engines, is well known.

Today's engine manufacturers require that the head gasket perform without a retorque operation, seal for extensive periods of time (750,000 miles or more in some highway truck cases) and "come off clean" so no scraping of the mating flanges is necessary when the engine is repaired. It sometimes needs to possess very high thermal conductivity to efficiently transfer heat between the block and head. It must be of such construction to permit rough handling and to have extended storage life.

The gasket must also perform in temperature ranges which are well below freezing at start-up to over 600°F in the combustion seal area during engine operation. It must accept occasional cases of detonation without failure. This is especially true today when premium fuels are sometimes unavailable and detonation associated with regular or no lead gas is possible.

The gasket must typically withstand combustion pressures of 1000 psi in naturally aspirated spark-ignition engines and 2700 psi or higher in turbocharged diesel applications. Today's gasket must also accommodate greater motions, both thermal and mechanical, as lighter-weight castings and lighter-weight, less rigid materials are being utilized for cylinder heads and engine blocks. As a result of the above requirements, the head gasket is a complex product.

Bolted joint and dynamic gasket sealing theory

Another reason why the head gasket is complex is the elasticities of the various components in the cylinder head–cylinder block bolted joint. Figure 10.1 depicts an elastic deflection schematic of this environment. Note that the head gasket is but one of several components in the total elastic deflection system. Owing to the numerous elastic sections of the system, the minimum sealing stress level on the gasket at the time of combustion is difficult to determine. This is the stress level on the gasket at the time of greatest sealing need and must be high enough for effective sealing.

Figure 10.2 depicts, in simple form, the forces and deflections in a typical head gasket joint.

Point A is the initial sealing point at clamp-up. The clamp load (C_{L_1}) equals the gasket load (G_{L_1}). Line OA is the stretch in the fastener and is determined by tension and extension testing of the bolts. Line AB is the elastic compression of the system, a portion of which is the elastic deflection of the gasket. At ignition there is a dynamic increase in the bolt load to point C_{L_2} and a decrease in the gasket load to point G_{L_2}. The numerical sum of these loads is equal to the product of the firing pressure and the area of the cylinder bore (FC = force of combustion). Point C_{L_2} is usually determined from the strain-gauged bolt data and G_{L_2} is normally determined by subtracting the combustion force value from point C_{L_2}. G_{L_2} is the critical value on this figure.

If the gasket stress associated with this point is adequate, sealing will result. If it is less than the minimum stress required for any given gasket design, leakage will result. One theory associated with the required minimum sealing is that it be high enough to create frictional forces on the gasket adequate in magnitude to resist radial motion of the gasket due to the combustion force. Eliminating this micromotion will result in maintaining the initial clamp-up sealing condition. Micromotion can result in localized fretting and possible cre-

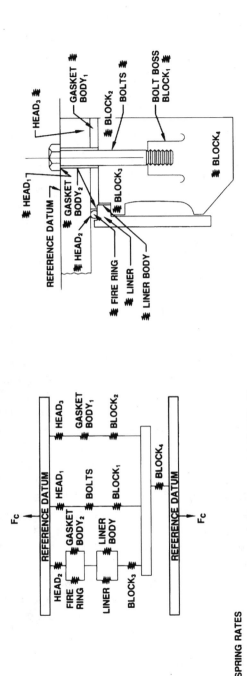

Figure 10.1 Elastic deflection schematic of the head gasket environment.

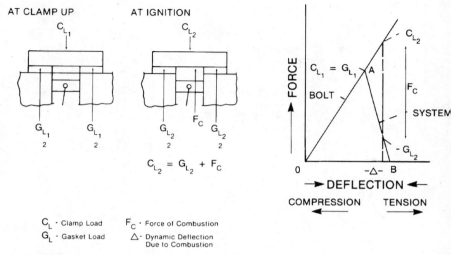

Figure 10.2 Forces and deflections in a typical head gasket joint.

ation of leakage paths. Experience indicates that for today's engines, the clamping load of the cylinder head bolts should be greater than 2½ times the combustion force for naturally aspirated spark ignition engines. It should be at least 3 times the combustion force for turbocharged diesel engines. The ratios are based on the total clamping force of the cylinder head and the total hydrostatic end force of all the cylinders. If these ratios are met or exceeded, the minimum sealing stress on the gasket is generally high enough to permit effective sealing. There are exceptions, of course, but the creation of these ratios during engine design is recommended. Although the above is a simple look at a complex situation, it is the basic theory involved in dynamic sealing, and consideration of the concepts involved results in improved gasket designs.

Because of the complex environment and extensive testing that engine gaskets are associated with, a design that seals a given engine is but one solution. That is, it is a design that works but is not necessarily the best or only design that has acceptable performance level for sealing the particular engine. Figure 10.3 graphically describes this.

The best performer is design H, but designs F, G, I, J, and K are also acceptable for the particular application. Once an

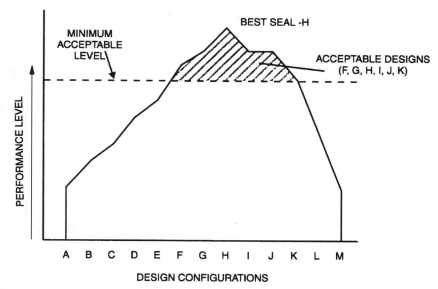

Figure 10.3 Performance level for various gasket designs.

engine manufacturer has tested and adopted one of the acceptable designs, pursuit of best or optimum design is not normally continued because of the high cost of testing involved.

History of gaskets and engines

Gasketing of the internal combustion engine has largely been one of reaction in that the sophistication of the gaskets has followed the extent of the sophistication in the engine itself. That is, as the engine became more powerful and placed greater demands on the gasketing, the gasket industry responded by designing the required sophistication into the gaskets. Figure 10.4 depicts the sophistication of engines and head gaskets versus time. The various time zones shown in this figure are associated with the specific information as shown in Table 10.2.

Head gaskets—time zones

Zone A. Initially, the basic head gaskets for the engines consisted of the sandwich type with asbestos millboard center, and

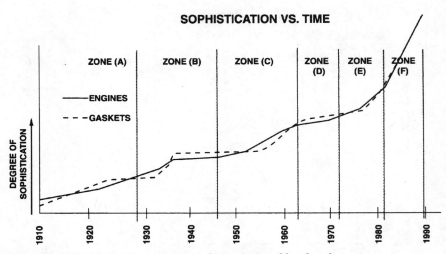

Figure 10.4 Sophistication versus time for engines and head gaskets.

TABLE 10.2   Head Gaskets Used in Various Time Zones.

HEAD GASKET DESIGNS VS. ENGINE SOPHISTICATION

ZONE	ENGINE SOPHISTICATION	HEAD GASKETS
A	Low output engines and Low compression ratios	Sandwich construction containing copper - asbestos - copper and gasket shellac; also, resin dipped asbestos millboard
B	Additional Cylinders added for more power In-Line Eights & Twelves Slight increase in compression ratio	Steel - asbestos - copper Steel used in combustion area for increased life
C	New Vee and OHV engines	Embossed steel with pre-applied coatings Various sealing aids in sandwich gaskets
D	High performance, large bore engines and automatic transmissions No retorque of cylinder head bolts	Development of bitumastic and rubber-fiber beater sheets used with tanged core Improved combustion seal design
E	Use of lighter weight alloys and castings and higher specific output CAFE & Clean Air Requirements	Development of laminated bodies and use of higher temperature resistant materials Improved anti-stick and anti-fret coatings Printing of elastomeric beads
F	More of E plus double overhead cam, more valves and turbocharging	Use of graphite, molded rubber & steel and multi-layer steel, rubber coated stainless steel materials

Engine Gaskets 295

either tinplate steel or copper used on the outer surfaces. Grommets and/or eyelets were incorporated in these gaskets depending upon the specific engine needs. Numerous versions were designed and manufactured. Figure 10.5 depicts a sandwich gasket.

Zone B. As the engines gained sophistication, the gaskets also gained in sophistication and a variety of designs were produced. These designs utilized various reinforcements at the combustion chamber seal for improved sealing. Metal shims and reinforced filler materials, for example, were incorporated in many constructions (Fig. 10.6).

Zone C. The embossed steel shim gasket was the next popular gasket to be utilized on passenger car engines (Fig. 10.7). This

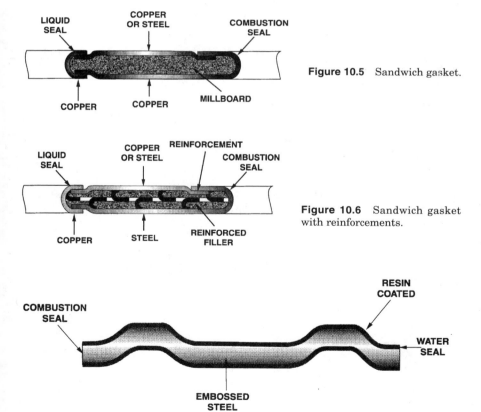

Figure 10.5 Sandwich gasket.

Figure 10.6 Sandwich gasket with reinforcements.

Figure 10.7 Embossed steel shim gasket.

gasket had a plastic resin coating applied for microsealing purposes. Because it was all metal, good torque retention was inherent with these gaskets. However, as engine displacement increased, the output resulted in motions that normally could not be accommodated by the elastic response of the embossed design. In addition, many times land areas, especially between cylinders, were reduced to a point where the legs of the emboss would fall inside the ports and adequate sealing was not possible.

Zone D. With the development of the rubber-fiber facing materials by the gasket paper manufacturers came improved designs. The majority of these designs incorporated a tanged or perforated core steel sheet, with these new facing materials mechanically clinched to either side of the core, thus providing soft surfaces of sealing material for water and oil sealing (Fig. 10.8). One of the major requirements of gaskets in this time zone was that they function without a retorque operation being conducted on the cylinder head bolts after the engine had been operated for some period of time.

Retorquing is essentially nonexistent in today's engines. While this may not appear to be a major requirement, it is indeed a major requirement since the retorque operation greatly aids gasket performance.

Zone E. One of the improved constructions eliminates the perforated core and uses an unbroken steel core to which an adhesive is applied for bonding the facings (Fig. 10.9). This laminated gasket has been adopted on many of the more difficult sealing applications. A recent technique used to seal critical passageways on today's engines is the utilization of silk screening to deposit elastomeric beads at these locations. In addition,

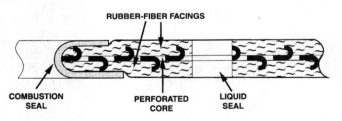

Figure 10.8 Perforated core head gasket.

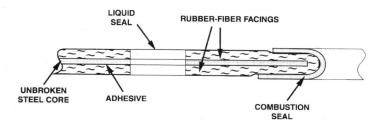

Figure 10.9 Unbroken metal core gasket.

many improvements in seal, antistick, and antifret coatings have been incorporated in the latest gasket constructions.

Zone F. Some of the latter designs have been a result of engine manufacturers' requirements for lower cylinder bore, cam bore, and valve seal distortions. These newer designs seal combustion at lower clamp loads, which in turn reduces these distortions. Figure 10.10 depicts one of these designs using multiple layers of steel and rubber coatings.

The sections that follow describe specific details of today's head gasket designs.

Head Gaskets—Combustion Sealing

In most cases a tinplate or zinc plate armor is used for sealing the combustion gases of spark ignition engines (Fig. 10.11). When an eyelet is used for combustion sealing of a head gasket, it is called by various names. At Fel-Pro it is called an armor. Other companies call it a flange or a bore eyelet.

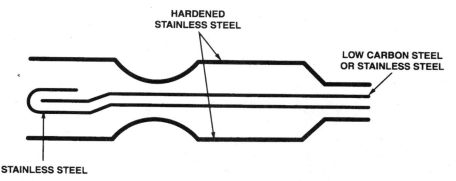

Figure 10.10 Multiple-layer steel gasket.

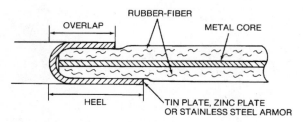

Figure 10.11 Armored gasket.

The thickness of the armor is a function of the thickness and type of facing material. The overlap and heel are sized for the specific engine to establish a proper unit seal load at the combustion chamber. The heel may be sized differently at various positions around the combustion chamber in order to obtain the proper unit loading at these positions (Fig. 10.12). High-output engines and/or turbocharged engines normally require stainless steel armor for improved high-temperature and fatigue resistance. Types 430, 304, and 321 stainless steel are commonly used.

In the case of diesel applications, an armored gasket is not generally adequate. Only certain thickness armors can be formed and embedded into given gasket bodies, and the thicknesses that normally fill these requirements are not structurally sufficient to withstand the high combustion pressures of these engines. As a result, other means for sealing combustion are necessary for these applications. The most popular method

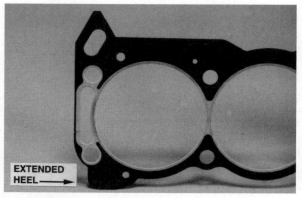

Figure 10.12 Extended heel.

incorporates the utilization of a low-carbon-steel ring. This ring gives a high unit sealing stress at very low loading and is widely used in today's diesel engines. The wire is butt-welded and generally attached to the gasket body by means of a stainless steel armor wrapping (Fig. 10.13).

In some cases, the wrapping may be tabbed to reduce the load required to embed the armor into the body, thus increasing the loading on the wire ring. In some cases, stainless-steel wires are necessary to withstand the heat and fatigue characteristics inherent in particular engines. An example is the case where precombustion chambers experience high thermal and mechanical movements. In some gaskets more than one wire may be utilized to achieve the desired sealing requirements of the engine.

Armored embossed metal is also used to seal combustion in a number of engines (Fig. 10.14). Varying the height and/or width of the emboss results in a wide range of load-compression properties. When the embossing is made from the core of the gasket body, thickness tolerance variations are minimized since the emboss and the core are made from the same piece of metal. Stainless or low-carbon steel are used as armors.

Diesel engines frequently have wet liners and the gasket is usually charged with sealing the intersection of the liner and crankcase (Fig. 10.15). During engine operation, there is

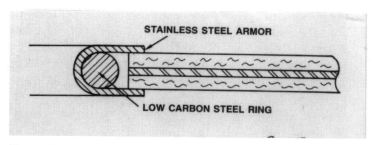

Figure 10.13 Wire ring gasket.

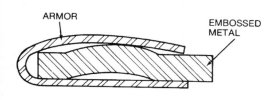

Figure 10.14 Armored embossed metal gasket.

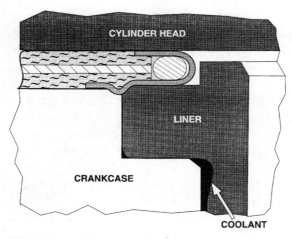

Figure 10.15 Wet liner engine.

motion between the liner and the block, and the likelihood of erosion of the liner seat is high. As a result, coolant can leak to the top of the deck and the gasket is required to seal at this location. In some cases, the soft surface is used to seal, while in other cases, the heel of the armor is extended to cover the intersection for sealing purposes. Engine testing normally dictates which is best. Some manufacturers are using room temperature vulcanizing (RTV) silicone to seal this application.

Other engines have liner designs that incorporate ridges, and in some cases grooves, in the cylinder head. This results in coining or embedding of the gasket for improved combustion gas sealing. Figure 10.16 shows one of these types.

Another head gasket design used on ridged liner engines is a thick 0.080-in steel plate which is embossed for improved combustion sealing (Fig. 10.17).

A unique gasket, which is used to seal very large diesel engines, uses copper-clad steel that has been etched away at various locations. The etching removes the copper from specific areas, thereby permitting high unit loading at other locations for improved combustion gas sealing (Fig. 10.18).

One of the items to be considered in the design of combustion sealing concepts is bore distortion. Some of the designs may need supplementary aids to keep bore distortion within acceptable limits. A few of the techniques used in this regard include

Engine Gaskets

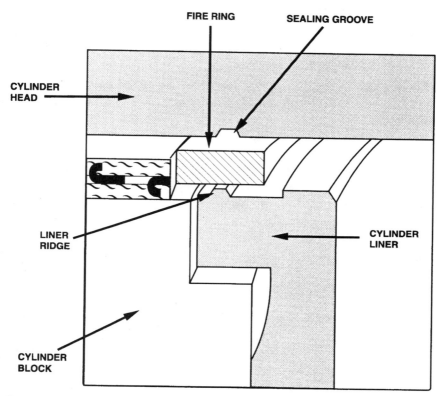

Figure 10.16 Ridge and groove gasket.

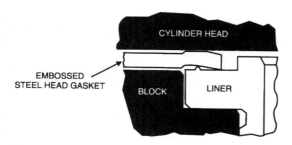

Figure 10.17 Thick steel plate gasket.

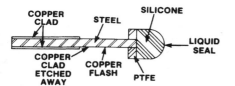

Figure 10.18 Etched copper clad gasket.

extending the combustion armor heel at specific locations (called lacework); overlapping the heel around the gasket body, generally at the ends of the gasket; and depositing beads or areas of rigid materials at preselected points. All these techniques essentially change the load-transmitting characteristics of the gasket and are useful for minimizing head bending as well as reducing bore distortion.

In some engines, the back-to-back location of exhaust valves results in high thermal growth in the area between cylinder bores. If excessive, this growth can result in combustion leakage. One means of improving the sealing in this area is to incorporate a metal shim in the gasket at this location. The shim acts as a stopper, permitting the gasket to resist the thermal growth and enhance sealing.

Air-cooled engines have somewhat reduced requirements in regard to head gasket sealing. Since there are no cooling water passageways, slight combustion gas leakage can be permitted as long as (1) engine performance is not affected and/or (2) the gasket is not affected by the leakage. Most of the gaskets for these engines consist of metal tanged core on both outer surfaces and a high-temperature-resistant fibrous core material. Because these engines are mainly made from aluminum, high thermal motions occur. The metal surfaces of the gasket permit head and block motions to occur without serious effect to the gasket's sealing ability (Fig. 10.19). Embossed metal gaskets are also used on these engines, especially when high heat transfer through the gasket is required.

Head Gaskets—Liquid Sealing

The basic factor involved in the creation and maintaining of the liquid seal is to have sufficient sealing stress on the gasket to

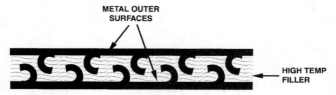

Figure 10.19 Metal-clad gasket.

assure conformation of the gasket to flange surfaces. This results in blocking the passage of media between the gasket and the flanges. In addition, the stress must be high enough to close any voids in the base material. The stress, however, must be low enough not to result in extrusion of the base material. To ensure long-term sealing, the material must, of course, retain adequate stress. Therefore, the selection of the facing material and its thickness is critical.

The base materials used for the gasket bodies must be extensively evaluated. The various ASTM test specifications for the materials' physical properties before and after fluid immersion and heat aging are conducted. In addition, bench test results for sealability, creep relaxation, crush and extrusion, etc., are analyzed before the material is accepted for use in a head gasket.

Some of today's head gaskets utilize soft surfaces for sealing the engine's liquids. These surfaces are rubber-fiber facings which are attached either mechanically and/or chemically to a metal core. The most popular facings use nitrile, neoprene, or polyacrylic elastomers. They are compounded to resist degradation by the oils and coolants, retain torque, minimize extrusion, and exhibit heat resistance. In addition, they must permit coolant infringement on the surfaces without degradation since the gaskets are used to meter and/or block coolant flow in many engines.

The mechanically clinched design uses perforated metal which has tangs on each side to which the facing is mechanically attached. They, in general, give adequate performance, although at times, in critical sealing areas, the liquids tend to seep along the core tangs, resulting in some leakage problems. In addition, if the tangs penetrate the outer surfaces of the gasket, erosion and/or corrosion of the mating flanges may occur. This is particularly true in the case of neoprene bond facing materials. The neoprene on aging may release hydrochloric acid, causing corrosion and etching.

The laminated or chemically bonded gasket bodies utilize an unbroken metal core to which the facings are bonded. Since there are no tangs, there is no possibility of leakage around the tangs or etching of the mating surfaces. The bonding adhesives, however, must be carefully compounded to accept the heat of the application and be resisted by a high-strength, facing-adhesive–core bond. As mentioned previously, these bodies may be embossed to achieve high unit sealing stress at various locations. In some cases metallic eyelets may be utilized at high-pressure openings to improve the sealing efficiency of the gasket at these locations.

This type of design lends itself to providing multiple-thickness gaskets. The steel core is varied in thickness while the amount of facing material is kept constant from gasket to gasket. This results in essentially equivalent torque retention properties even though the gaskets differ in thickness. The gaskets are desired to have different thicknesses to control compression rates. Some production tolerances of engine parts and some cylinder head machining in service will result in lack of compression ratio control. Using different thicknesses of head gaskets compensates for these. The combustion seal is varied to compensate for the change in the steel core (Fig. 10.20).

A number of thermosetting seal and antistick coatings are used on gasket bodies. They provide microsealing properties to the gasket and eliminate sticking of the facing to the head or block when the engine is disassembled.

Some gasket manufacturers impregnate various gasket bodies with various reasons for improved sealing properties. The impregnation is usually associated with bitumastic bound sheets, which are rarely used in the United States today.

An important physical property of a gasket body is good torque retention properties. In general, torque retention is

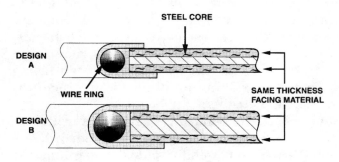

Figure 10.20 Multiple-thickness gaskets.

associated with the amount of compressible material in a gasket. Compression set is a function of the compressible amount, and reduction of the compressible amount results in higher retained torque of the fasteners. There must be sufficient compressibility consistent with good seal, however. Some of the newer Teflon coatings, which will fill in surface irregularities in cylinder blocks and cylinder heads, permit the reduction of the amount of compressible material in a given gasket.

Antifret coatings are used to keep the gasket and/or mating flanges from fretting due to the motions that occur during engine firing. Figure 10.21 shows some of the motions that occur as the engine is operated. Molybdenum disulfide, graphite, and Teflon are some of the compounds used in these coatings. Today's new lightweight engines are resulting in larger motions and the antifret coatings are becoming more prevalent in today's gasket designs.

In cases of large head-to-block motion that may occur at specific liquid passageways, molded rubber, either with or without metal reinforcement, is sometimes incorporated in the gasket for accommodation of the motion and effective sealing. These grommets are either bonded and/or staked in place, depending upon the gasket design (Fig. 10.22).

A popular technique utilized to improve the liquid sealing of head gaskets is the printing or silk screening of elastomeric beads at the liquid ports. This technique was described in detail earlier. When a thin-cored gasket body is used, the beads are usually located on one surface of the gasket, generally on the surface facing the weaker mating flange. The thin core allows transfer of the localized stress through the gasket to the opposite flange. The beads may be positioned on both sides of the gasket. One example is when the steel core is of substantial thickness and the transfer of a high unit load is impossible because of the thick core. A variety of materials can be utilized in this technique, the most popular being silicone. The thickness of the bead is somewhat dependent upon the nature, type, and thickness of the facing to which it is applied. Normally, as the facing material becomes thicker, a thicker bead is deposited.

In-line engines many times present difficult liquid sealing problems. They are inherently unbalanced in regard to distribution of bolt loading. In many cases, no studs or cap screws are

306 Chapter Ten

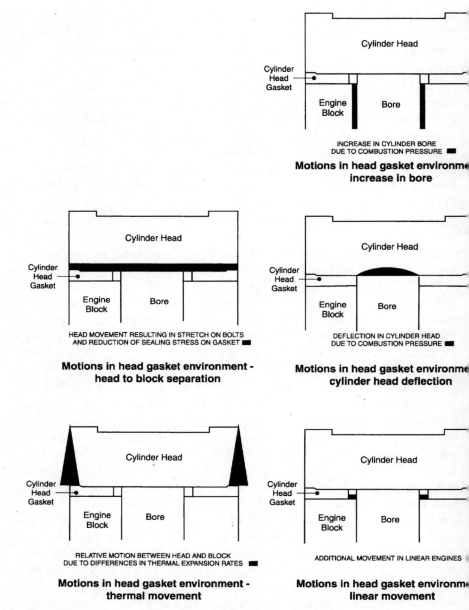

Figure 10.21 Various motions in the head gasket environment.

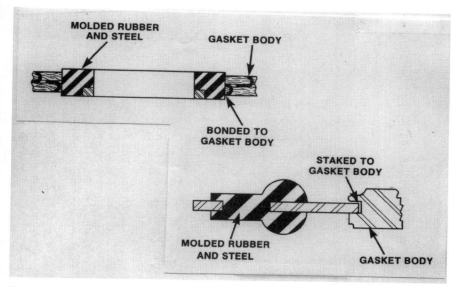

Figure 10.22 Molded rubber grommets used in head gaskets.

used to provide clamping pressure between head and block on the outer periphery of the push rod cavity. Supplementary sealing means must be designed into the gasket; otherwise oil seepage will occur down the side of the block. Two gasket design techniques are commonly used to solve this problem. They are:

1. Dipping the push rod area of the gasket to give this area a rubber overcoat
2. Applying a bead of high-temperature synthetic rubber to one or both faces of the gasket around the low clamp load area

In each of these cases the intent is to obtain adequate clamp load at the push rod area by building up the gasket thickness by means of the rubber overcoat or the silk-screened bead.

The open deck or open tank engines, which are becoming more popular because of their light weight, pose additional sealing demands on gasket bodies. These engine designs result in large water and steam impingement sections on the gasket. These areas are unsupported and are undergoing pressure pulsations as the engine operates. Sealing the fluids, being resistant to degradation by them, and withstanding their corrosive attack are requirements of the gasket bodies on these engines. Providing small steam vent holes in the gasket where possible is recommended.

During the last few years, multilayer metal gaskets incorporating rubber coatings have been the designs of choice for many engine manufacturers. These gaskets are normally three or four layers and utilize embossed, full-hard, stainless-steel metals. The metals are coated with either nitrile or viton coatings with at least one coating between any two metals. The gaskets exhibit very high recovery and seal combustion at low unit loadings.

Thick steel plates (0.080 in) with edge molded rubber for liquid sealing are also being adopted, especially for one- or two-cylinder gaskets for heavy-duty diesel engines.

Intake and Exhaust Manifold Gaskets

The remaining metal gaskets on engines are associated with the intake and exhaust manifolds. Intake manifold gaskets must have excellent vacuum sealing ability because even slight air leakage increases the air-fuel ratio, which may result in a hot running engine, poor idle, increase in NO_x pollution and probable reduction of valve life. Also, the computer control of engine operation demands high-performance sealing or engine operation becomes very unstable. In addition to sealing vacuum, many intake manifold gaskets have water crossover ports, and on V engines, many of them have exhaust gas crossover ports for heating the fuel charge.

Sealing of these gaskets follows the same general criteria as the cylinder head gasket bodies, as discussed above. One of the major differences is the generally lower clamping load existent on these applications. Because of the light clamping load, many soft gaskets without metal cores were not adequately compressed to stop wicking of coolant and fuel vapors. As a result, they tended to break down and rupture in the port walls, which resulted in either vacuum or water leaks. The adhesively bonded laminate construction containing a metal core is many times used for these applications because of its inherent radial strength and ability to exhibit high unit sealing stresses if embossed.

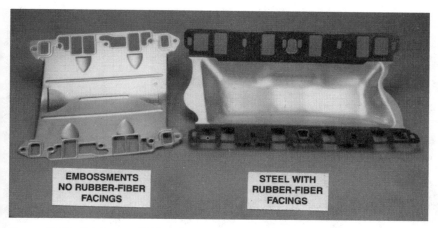

Figure 10.23 Intake manifold gasket with combined oil splash plate.

Some V production engines have incorporated intake manifold gaskets with an integral oil splash plate which stops oil splash on the base of the manifold (Fig. 10.23). This is done to prevent oil sludge from caking on the hot section of the manifold which is porting the exhaust gas crossover. In these cases, the metal splash plate may have rubber fiber facings attached to it for improved sealing of the manifold section. Again, these ports may see embossing utilized for improved sealing. In some cases, a metal eyelet may be used in the crossover port.

Many spark ignition engines do not need exhaust manifold gaskets in the production build. Service engines normally do need them, however, owing to thermal warpage which occurs after the engine is operated. Generally speaking, the exhaust gasket contains a steel surface at least on the side facing the manifold to permit manifold-to-gasket slippage. If the gasket has an overlap, it is located on the stationary side of the joint. To accommodate large thermal motion that occurs on some manifolds, some gaskets have slotted bolt holes and some gaskets may have a crimp incorporated in them (Fig. 10.24).

Often single-port, embossed steel gaskets are used on heavy-duty diesel engines for better accommodation of the large thermal motion that occurs. In many cases, single-bead embossments are adequate but some assemblies require a double-bead design (Fig. 10.25).

Figure 10.24 Exhaust manifold gasket with crimp.

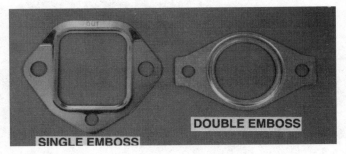

Figure 10.25 Single- and double-embossed gaskets.

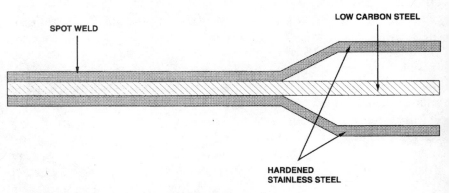

Figure 10.26 Multiple-layer embossed exhaust gasket.

To accommodate even larger motions, multiple layer embossed gaskets can be utilized (see Fig. 10.26). To further aid in cases where large motion is involved, these gaskets can be coated with high-temperature solid film lubricants such as graphite, molybdenum disulfide, or other special formulations.

Some exhaust manifold gaskets also incorporate additional material to act as a heat shield to protect either ignition wires

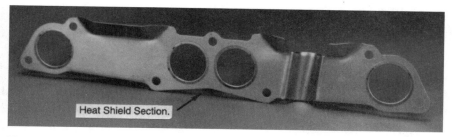

Figure 10.27 Exhaust manifold gasket with heat shield—Example 1.

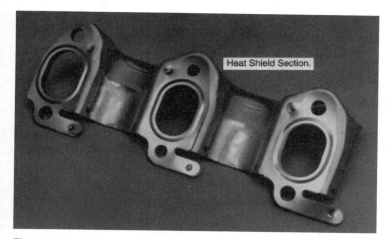

Figure 10.28 Exhaust manifold gasket with heat shield—Example 2.

and/or other engine rubber parts from the high heat of the engine exhaust. Figures 10.27 and 10.28 show two of these.

Another type of seal which is external to the engine, but one which is associated with sealing exhaust gas, is the exhaust pipe ring seal located between the exhaust header and the tailpipe. These gaskets are normally made from bitumastic or cement bound fibrous material with metal reinforcement. In some cases, where very high temperatures are experienced, a metal sleeve, often stainless, is incorporated in the seal to protect the material from eroding at high temperatures, which results in long-term sealing. This particular gasket is in an even more demanding environment in transverse-mounted engines. Rotation of the engine during operation causes motion on the sealing surfaces. The gaskets in these engines must accommodate this motion while continuing to seal.

Other Gaskets

Basically, the other gaskets on the engine fall into two general categories. They are attached to the engine either with stamped sheet-metal covers or with cast housings, and the gaskets used differ depending upon the attaching cover. For example, the stamped metal covers are easily distorted around the bolt holes by overtorquing while the cast covers take considerable load to distort. In some of the latter cases, center hold-down bolts are utilized.

Because of the metal distortion problem, softer and thicker materials have generally been selected for the stamped metal cover applications. Rocker and cam covers, push rod covers, and oil pans of spark ignition and smaller diesel engines are normally of the stamped metal design. The older gaskets for these applications were generally resin-bound cork. Later gaskets were usually made out of synthetic rubber and/or cork with rubber binders, replacing the moisture-sensitive and dimensionally unstable resin binder. In some cases, rubber-coated organic fiber materials were also used for these applications. These gaskets are normally not rigid, and when used on nonflat surfaces, such as cylinder heads, the gaskets may move out of position during installation. One technique used to aid in these cases is to employ a metal core in the gasket to give it stiffness. The core also aids in avoiding crushing of the gasket in the bolt hole locations since less deflection per unit load occurs with a cored gasket.

Newer gaskets for these applications are molded rubber, molded rubber and metal, or molded rubber and plastic. Silicone rubber is the most popular polymer.

Most heavy-duty engines utilize cast housings for the oil pan and sometimes for the rocker cover. A wide variety of rubber-fiber-type gaskets are used in these cases. These castings generally have better clamping loads and more rigid flanges than sheet-metal covers. Sometimes, however, bolt spacing on these is large and inadequate gasket clamping pressure between the bolts is manifest.

Improvements in gaskets, such as printed elastomeric beads, discussed earlier, may be utilized in these cases. Thicker materials are not normally selected for these applications for improved sealing because, as the material gets thicker and the compressibility is increased, the tendency for torque loss is increased. This torque loss can result in long-term sealing problems.

Timing cover gaskets and associated metal parts essentially follow the same guidelines as above, depending upon the cover design.

Some of the newer laws that affect the gasketing of internal combustion engines are noise pollution and vibration. This is called NVH (noise and vibration harshness). At the current time, many engine manufacturers are looking into the possibility of reducing noise from the various cover areas of different engines. Plastic and fiberglass-reinforced plastic are being considered for potential adoption for some of these covers. Presently, the trend in gasketing for these applications is concentrated on thick rubber gaskets for vibration isolation purposes as well as long-term sealing.

Although all U.S.-made engines of today are fuel-injected, there are carbureted engines on the road. Carburetor base gaskets must be designed to keep evaporative emissions minimized and provide various insulating properties depending upon the engine requirements.

Accessory items, such as oil pump and fuel pump assemblies, require a specific type of gasket material. For these functions, power must be transmitted through the flange joint by either a rotating shaft or lever; thus the flange joint is subjected to a working load. Gasket materials for these applications are generally fiber and are of higher density. These are specified to keep torque loss at a minimum and therefore reduce the possibility of the bolts vibrating loose and causing eventual leakage. In some cases, such as pumps, the gasket also functions as a shim.

Chemical Gasketing

A number of nonmetallic gasket applications are being sealed with chemical gaskets. They were discussed in detail earlier.

Use of chemical gaskets results in essentially metal-to-metal contact of the joint and therefore essentially zero torque loss. The sealing phenomenon behind these gaskets differs substantially from that of the mechanical gasket. The materials are not absorbing load, nor do they have any inherent compressibility in them. Because of this, they do not follow the mating flanges by uncompressing as they separate. Either the adhesion of the chemicals is strong enough to stop the motion or the chemical has adequate extensibility to accommodate the motion.

Chemicals have been used on engines for various applications such as rocker and cam covers, and water outlets and oil pans, and are growing in popularity. They are especially well suited for high-volume OE usages where automatic dispersing means can be afforded.

Service applications, however, differ from the production build and must be considered independently for possible use of chemical gasketing. Some service applications may use a mechanical gasket even thought the OE build used a chemical gasket. This may be due, for example, to the difficulty in applying the correct amount of the chemical in the servicing of the engine or may be due to the time element involved in assembling the flanges while the chemical is uncured.

There is a possibility that one may see one gasket for the OE build and a different design for service; i.e., the OE gasket may be a chemical gasket applied with automatic dispensing equipment while the service gasket may be a mechanical gasket installed under the hood of a vehicle. One thing for certain is that the gasket industry recognizes the difference between the production and service requirements and must respond to both.

Testing of Engine Gaskets

Bore distortion

One important characteristic associated with head gasket sealing is associated with the engine's cylinder bore distortion. As the gasket is clamped, the loading on the engine block results in distortion of the metal on the cylinder bores. Several thousandths of an inch can occur. The thin piston rings used on today's engines for reduced friction and improved fuel economy require round cylinder bores. Therefore, engine manufacturers are specifying a maximum amount of bore distortion.

A sophisticated instrument is available to measure bore distortion. It is called an Incometer. It is used to measure cylinder bore roundness and vertical uniformity under ambient static conditions.

This equipment provides measurements of the initial "free state" cylinder bore as compared to the cylinder bore after a gasket has been installed and the cylinder head torqued to the required level. This allows gasket designs to be compared and evaluated for their effects on cylinder bore distortion.

Minimizing cylinder bore distortion improves piston ring to cylinder bore contact, resulting in reduced oil consumption, exhaust emissions, and piston ring blow-by. Figure 10.29 shows the bore distortion test unit, and Fig. 10.30 is a close-up of the unit in an engine's cylinder bore.

Figure 10.31 is a radial plot of bore distortion. Figure 10.32 is a three-dimensional plot of bore distortion.

Figure 10.29 Bore distortion test unit.

Figure 10.30 Close-up of an engine's cylinder bore.

Vacuum testing

Vacuum (negative gauge pressure) testing is used to evaluate the sealing performance of gaskets using vacuum within the cooling cavities of the engine assembly. The required testing is accomplished using a high-capacity vacuum pump with the necessary manifold and vacuum gauges. After the required level of vacuum is achieved, the valves are turned to OFF position and vacuum decay is monitored over time. The use of vacuum provides a method to monitor and record actual values for gasket sealing performance. This type of testing has been especially beneficial in evaluating materials, coatings, and designs of cylinder head and intake manifold gaskets for sealing performance using various mating flange surface finishes. Figure 10.33 shows an engine being vacuum tested.

Nitrogen gas combustion testing

High-pressure nitrogen gas is used to evaluate cylinder head gasket combustion seal performance. Inexpensive and readily available in 2000 psig cylinder bottles, high-pressure nitrogen

Engine Gaskets 317

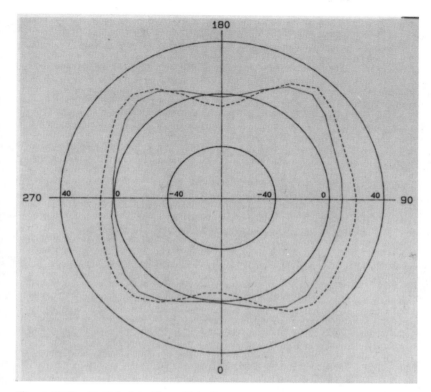

Figure 10.31 Radial plot of bore distortion.

gas is used to simulate combustion gas pressure(s) within the engine cylinders. The resulting gathered information provides cylinder head gasket combustion seal sealing performance between cylinders, to the engine coolant ports, and to the outside.

General test parameters

Standard test pressures	500 psig
	800 psig
	1000 psig (max)
Sealing performance monitor locations	Cylinder to coolant
	Cylinder to cylinder
	Cylinder to outside

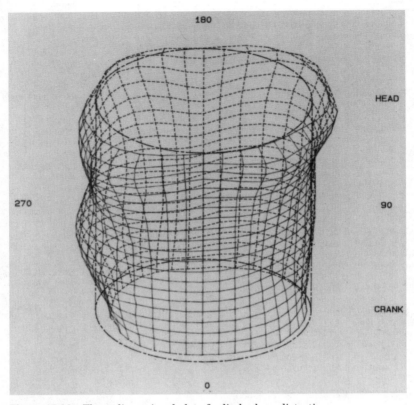

Figure 10.32 Three-dimensional plot of cylinder bore distortion.

Test temperature   Ambient (at 70°F)
Elevated temperatures may be requested

Figure 10.34 shows an engine being nitrogen tested.

Engine coolant testing

Engine coolant sealability testing is used for evaluating gasket fluid sealing performances on an actual engine. The engine is used as the test fixture, filled with coolant, and pressurized. Gasket sealing performance is evaluated with the aid of soap solution (bubbles) and/or a black light. This type of testing has been mostly used to evaluate the sealing performance of cylinder head and intake manifold (coolant crossover) gaskets.

Engine Gaskets 319

Figure 10.33 Engine being vacuum tested.

Figure 10.34 Nitrogen gas combustion sealability testing.

General test parameters

Coolant (std)	50/50 percent glycol base antifreeze and water
Air pressure applied to coolant (10 psig increments)	60 psig (std max) 120 psig (system max)
Fluorescent dye	Added to coolant

Figure 10.35 shows engine coolant testing.

Head-to-block motion testing

Capacitance probe. Capacitance probes are noncontacting displacement measuring transducers. They are used to measure and record the relative motion or change in position of metallic surfaces. These probes have been used to measure two-dimensional thermal growth of cylinder heads and the relative motion between a cylinder head and cylinder block. Figure 10.36 depicts capacitance probe displacement measurement testing.

Gap sensor. Gap sensor can be used to measure and record the distance or change in distance between two metallic surfaces such as a gasketed joint gap. The system uses small ($1/4$-in diameter) bidirectional eddy-current sensors (transducers) that are installed into prepunched holes in the gasket for static and/or dynamic testing on an actual engine.

Static testing

The system has been mainly used to measure engine cylinder head lift during intake manifold installation. Figure 10.37 depicts a gap sensor in a gasket.

Figure 10.38 shows the measuring instrumentation used in gap sensor testing. Referenced SAE Paper 890272 discusses gap sensor testing and test results.

Dynamometer testing

Engine testing (gasoline and diesel) is used to evaluate gasket performance under actual operating conditions. Testing is accomplished using dynamometer-equipped test cells and field

Engine Gaskets 321

Figure 10.35 Engine coolant testing with fluorescent die and coolant/oil.

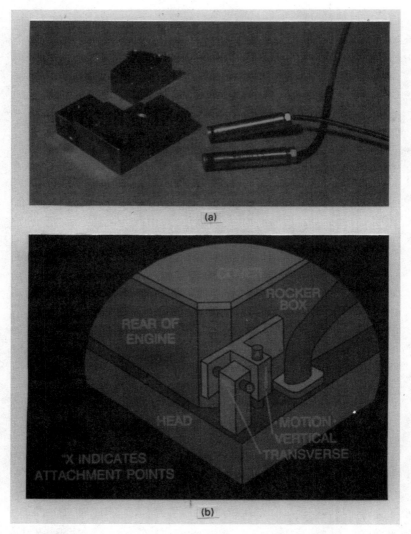

Figure 10.36 Two capacitance probes and use of the system in an engine application.

test vehicles. Dynamometer testing provides gasket performance information under controlled test conditions. Figure 10.39 depicts an engine dynamometer and data acquisition instrumentation.

Hydrocarbon leak detection can be used to evaluate cylinder head gasket combustion seal performance during dynamometer

Engine Gaskets 323

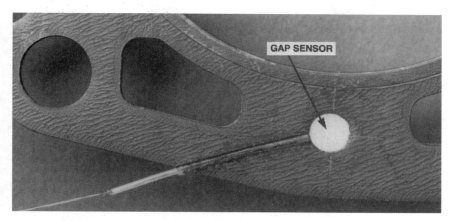

Figure 10.37 Gap sensor located in a gasket.

Figure 10.38 Measuring instrumentation used in gap sensor testing.

testing. The method incorporates the collection of a gas sample from the engine cooling system expansion tank and analyzing the sample for hydrocarbon content level in parts per million using gas chromatography.

Test results have proved this method provides good indications of combustion seal performance as observed by combus-

Figure 10.39 Engine dynamometer testing and data acquisition instrumentation.

tion leakage into the cooling system. This test method has been especially useful in evaluating cylinder head gasket sealing characteristics as gasket load is decreased (cylinder head bolt torque reduced).

Test results using this test method have shown good correlation with the results using high-pressure nitrogen gas combustion seal testing. Experience indicates that this test method is best used as a comparative tool in evaluating combustion seal performance, that is, monitoring hydrocarbon level change (increase) throughout the engine test run.

Field testing

Field testing of engine gaskets on operating engines is normally conducted prior to approval of the gaskets for production.

References

"Additional Guidelines for Internal Combustion Engine Gaskets—Liquid Sealing," A-13 Supplement A, 1993.

Crouch, L. E., D. E. Czernik, V. J. Labrecque, and S. M. Lillis: "A Well Designed Cylinder Head Gasket—That Thin Margin of Success," SAE paper 771B, October 1963.

Crowe, J. R.: "Mack's Fire Ring Gasket," SAE paper 700026, January 1970.

Czernik, D. E., J. C. Moerk, Jr., and F. A. Robbins: "The Relationship of a Gasket's Physical Properties to the Sealing Phenomenon." SAE paper 650431, May 1965.

Czernik, D. E.: "Sealing Today's Engines," *Fleet Maintenance & Specifying,* Irving Cloud Publishers, July 1977.

"Gasket and Joint Design Manual for Engine and Transmission Systems," SAE publication AE-13, 1988.

Lillis, S. M.: "Getting to the Core of Gasket Sealing," SAE paper 680027, January 1968.

McDowell, Donald J.: "Choose the Right Gasket Material," *Assembly Engineering,* Hitchcock Publishing Co., October 1978.

Petrunich, P. S. "Gasket Designs and Applications Using Flexible Graphite." SAE paper 830214 presented at the SAE International Congress and Exposition, February 1983.

Bibliography of SAE papers associated with head gaskets

Albertz, K. M., and K. G. Freidrichs: "Cylinder Head Gaskets for Bimetal Engines and Techniques for Optimizing the Method of Clamping," SAE paper 810365, presented at the SAE International Congress and Exposition, February 1981.

Berner, Dietrick: "Non-Asbestos and Asbestos Based Cylinder Head Gaskets: Differences in the Basic Design Principles," SAE paper 870008, presented at the SAE International Congress and Exposition, February 1987.

Choi, Paul S., John A. Damusis, and Richard J. Kozerski: "The Effect of CAD/CAM on Gasket Design and Manufacturing," SAE paper 860629, presented at the SAE International Congress and Exposition, February 1986.

Connelly, Terrence T., and Jeffrey E. Spencer: "Automated Control System for Thermal Shock Engine Testing," SAE paper 880141, presented at the SAE International Congress and Exposition, February 1988.

Crowe, J. R.: "Mack's Fire Ring Gasket," SAE paper 700026, presented at the Automotive Engineering Congress, January 1970.

Czernik, D. E., and F. L. Miszczak: "A New Technique to Measure Real Time Static and Dynamic Gasket Stresses," SAE paper 910205, 1991.

Czernik, D. E., J. C. Moerk, Jr., and F. A. Robbins: "The Relationship of a Gasket's Physical Properties to the Sealing Phenomenon," SAE paper 650431, presented at the Mid-Year Meeting, May 1965.

Czernik, D., B. Pearlstein, and J. Amodeo: "Investigation of the Cylinder Head to Crankcase Motions on a Specific Bi-Metal Engine," SAE paper 890272, 1989.

Czernik, Daniel E., and Gerald Rosenquist: "Gasketing the High Performance Engine," SAE paper 850187, presented at the SAE International Congress and Exposition, February 1985.

Czernik, Daniel E., and Robert F. Kovarik: "The Combustion Seal: Theory and Performance," SAE paper 870006, presented at the SAE International Congress and Exposition, February 1987.

Czernik, Daniel E.: "Designing and Testing Cylinder Head Gaskets for High Performance Sealing," SAE paper 840188, presented at the SAE International Congress and Exposition, February 1984.

Dishaw, R. D., and M. Kobayashi: "Multi-Layer Steel Gasket Technology for Engine Cylinder Head and Manifold Applications," SAE paper 890270, 1989.

Fell, Gary C., and Robert F. Kovarik, Jr.: "Test Methods for Predicting Engine Cylinder Head Gasket Performance," SAE paper 851565, presented at the SAE International off Highway and Powerplant Congress and Exposition, February 1985.

Finkelston, Robert O.: "The Effect of Bolt Tightening Characteristics and Gasket Properties on Cylinder Head Gasketing," SAE paper 810363, presented at the SAE International Congress and Exposition, February 1981.

Gronle, Hans-George: "Alternatives to Asbestos in Cylinder Head Gaskets," SAE paper 860627, presented at the SAE International Congress and Exposition, February 1986.

Grundler, R., and A. Rist: "Investigating the Static and Dynamic Loading of Cylinder Head Bolting," SAE paper 805148, presented at the 18th FISITA Congress, 1980.

Leonne, Klaus, and Ron Ziemba: "The GOETZE Cylinder Distortion Measurement System and the Possibilities of Reducing Cylinder Distortions," SAE paper 880142, presented at the SAE International Congress and Exposition, February 1988.

Lillis, S. M.: "Getting to the Core Gasket Sealing," SAE paper 680027, presented at the Automotive Engineering Congress, January 1968.

Lonne, Klaus, Klaus-Peter Majewski, and Brian A. Newman: "The Sliding Cylinder Head Gasket—A Solution for High-Duty Engines," SAE paper 840190, presented at the SAE International Congress and Exposition, February 1984.

Majewski, Klaus-Peter, Hans-Rainer Zerfass, and Mike Scislowicz: "Asbestos Substitution in Cylinder Head Gaskets," SAE paper 880144, presented at the SAE International Congress and Exposition, February 1988.

Nauman, Horst: "Cylinder Head Gasket Requirements for Opened and Closed Deck 390 Engines," SAE paper 830002, presented at the SAE International Congress and Exposition, February 1983.

Percival, Paul R., and Brian G. J. Williams: "Non-Asbestos Gasket Engineering," SAE paper 850191, presented at the SAE International Congress and Exposition, February 1985.

Pohle, Rainald: "Reinz—AFM 5, A New Asbestos-Free Cylinder Head Gasket Material," SAE paper 820141, presented at the SAE International Congress and Exposition, February 1982.

Russell, Richard G., and Terrence T. Connelly: "Thermal Shock Testing Head Gaskets," SAE paper 830212, presented at the SAE International Congress and Exposition, February 1983.

Sugawara, Minou: "Features of Expansion Graphite Cylinder Head Gaskets," SAE paper 850193, presented at the SAE International Congress and Exposition, February 1985.

Tensor, Paul M.: "Computer Aided Design and Load Analysis of a Cylinder Head Gasket," SAE paper 800071, presented at the SAE International Congress and Exposition, February 1980.

Teucher, G. S., and F. Stecher: "Cylinder Head Gaskets for High Peak Pressures," SAE paper 700025, presented at the Automotive Engineering Congress, January 1970.

Vaiden, R. E.: "Elastomeric Materials for Engine and Transmission Gaskets," SAE paper 921032, 1992.

Wagenplast, Dieter, and Hans-George Gronle: "High Density Composite—New Material for Non-Asbestos Gasketing," SAE paper 880138, presented at the SAE International Congress and Exposition, February 1988.

Widder, E. S., and G. J. Novak: "Gasketed Joint Analysis Using Computer Aided Engineering Techniques," SAE paper 920131, 1992.

Williams, B. G. J., and M. C. Bannard: "Cylinder Head Gasketing Problems in Bimetallic Engines," SAE paper 840189, presented at the SAE International Congress and Exposition, February 1984.

Yanagisawa, Takashi, and Masahiko Teramoto: "Optimum Design of Cylinder Head Gasket and Related Techniques," SAE paper 861376, presented at the SAE Passenger Car Meeting and Exposition, September 1986.

Yates, A., and P. N. W. Dudley: "Static Testing Methods for Evaluating Head Gasket Performance," SAE paper 830213, presented at the SAE International Congress and Exposition, February 1983.

Zeitz., J. E., J. A. Damusis, and J. A. Ulrich: "Designing with the New Asbestos-Free Gasket Materials," SAE paper 810366, presented at the SAE International Congress and Exposition, February 1981.

Index

"m" factors, 10
"Z" strength, 35

Acoustic, 3, 101
Acoustic isolation, 35
Acrylics, 38
Adhesion, 10
Adhesion bond strength, 10
AHOTT (Aged Hot Operational Tightness Test), 137
American Society of Mechanical Engineers (ASME), 2, 7, 79, 105, 125, 149
American Society of Test Methods (ASTM), 7
American Society of Testing Materials (ASTM), 1, 4, 7
Anaerobic, 238, 276–285
Antifret, 161
Antistick, 29, 35, 161
Antistick properties, 164, 230 (see Gasket application data sheet)
Antistick properties, 164
Armored gasket, 298
Asbestos, 4, 5, 7, 11, 13, 14, 18, 30–33, 37, 38, 45, 47, 49, 87, 127, 128, 134, 136, 139, 140, 142–145, 147, 149, 150, 229, 293, 325–327
ASME Code, 2, 24, 105, 108, 113–115, 117, 128, 131, 148
 Pressure Vessel Research Council (PVRC) simplified procedure—Whalen, 105, 111, 125
ASTM, 49, 72
ASTM F104 Nonmetallic Gasket Classification, 5

ASTM F868 Classification for Laminated Composite Gasket Materials, 11, 168
ASTM Standard Test Methods for gasket, 7
ASTM test methods for vulcanized elastomers, 10, 16, 199

Bench test, 246, 249, 303
 oven test, 248
 steam test, 148, 246, 247, 249, 250
 vibration table and environmental chamber test, 246
Binder, 3, 4, 11, 13, 27–32, 36–39, 41, 65, 82, 91, 134, 135, 168, 238, 312
Binder durability, 7
Binders, 4, 11, 28, 30–32, 36–39, 41, 135, 168, 312
Blowout, 25, 53, 55, 59, 68, 148
Blowout testing, 10
Bolt fastening sequence, 62

Carbonless paper, 70, 76–78, 82
Cellulose fiber, 238
Ceramics and various inorganic fibers, 38
CETIM, 25, 151
Chemical compatibility, 34
Chemical gasket, 125, 187, 276–279, 281–286, 313, 314
 cure properties, 278
 sealing properties, 12, 222, 304
 uncured properties, 278, 286
Chromatography, 239, 323
Clorosulphonated polyethylene, 39, 40
Coating, 161
Compressed gaskets, 261

329

Index

Compressed thickness, 36
Compressibility, 6, 7, 10, 16, 32, 34, 38, 41, 47, 52, 67–70, 179, 194, 230, 232, 282, 283, 305, 313, 314
Compression:
  deflection, 19
  limiter, 202
  molding, 200
  set, 12, 16, 40, 56, 156, 157, 191, 222, 232, 234, 285
Compression set, 10
Computers and gaskets, 81–83, 211, 219, 257, 262, 264, 308, 327
Conformability, 34, 44, 55, 66, 67, 203
Consistency of formulation, 35
Cork, 4, 5, 7, 11, 13, 14, 41, 49, 229, 312
Cork-rubber, 41
Corrosion testing, 10
Crack growth, 10
Creep relaxation, 7, 12, 36, 191, 221, 223, 231, 233
Crush resistance, 37, 177
Cure properties, 278
Cylinder head gasket, 289, 308, 316, 317, 322, 325, 326, 327

Densitometer, 78
Density, 35
Differential scanning calorimetry (DSC), 236
Differential thermal analysis (DTA), 236
Dimensional stability, 10, 16, 35
Dynamic mechanical analysis (DMA), 236

Ecole Polytechnique, 49, 147
Edge molded rubber gasket, 308
Elastomer, 22, 134, 138, 144–148, 151, 184, 185, 190, 196, 198, 199, 203, 205–208, 210, 219
Elastomeric beads, 180, 182, 184, 296, 305, 313
Electrical sensor, 70, 81, 82
Elevated Temperature Research Program, 133
Embossed steel shim gasket, 295
Embossing, 179, 299, 309
Epoxy/phenolic, 163
Erosion, 203, 300, 303
Erosion resistance, 34
Ethyelene, 39
Ethylene propylene, 39, 40

Extrusion, 31, 32, 35–39, 56, 174, 177, 213, 216–219, 222, 234, 303
Eyelet, 164, 243, 297, 309
Eyeletting, 164

Fastener, 55, 56, 66, 245, 250, 278, 290
  tightening, 59, 63, 76, 100, 125, 326
Fibers, 13, 14, 25, 27, 29–32, 36–39, 87, 127, 134, 229, 238
  asbestos, 4, 5, 11, 13, 14, 30–33, 38, 45, 47, 87, 127, 128, 134, 136, 139, 140, 142–145, 147, 149, 150, 229, 293, 326, 327
  inorganic, 4, 10, 11, 13–15, 31, 38, 41, 168, 229
  nonasbestos, 33, 38, 49
Filler, 39, 41, 46, 135, 196, 207, 208, 238, 295
  inert, 38, 41, 42, 206, 207, 237
  nonmetallic, 2, 4, 5, 10, 11, 13, 14, 42, 49, 64, 169, 223, 229–231, 233, 271
  reinforced, 25, 30, 32, 134, 138–140, 142, 144–146, 151, 168, 185, 295, 313
Finite element analysis (FEA), 260
Fire resistance tests, 136
Flange bending, 69, 229
Flange surface finish, 47, 113, 316
Flat metal gaskets, 47
Flexibility, 7
Fluid resistance, 7, 13, 18, 32, 37, 39, 227
Fluorocarbon (viton, technoflon, fluorel), 5, 40
Fluorosilicone, 39
Frictional coefficient gaskets, 66, 103
Fuji prescale film, 78

Gaskets, 1–4, 10, 11, 14, 15, 24, 25, 32, 33, 36, 38, 39, 41, 42, 44–48, 51, 54, 64, 65, 68–70, 104, 108, 114, 117, 123, 125, 127, 130–136, 138, 141–151, 160, 163, 164, 166, 168, 173, 177–180, 182, 183, 184, 186, 187, 189, 199, 202, 203, 210, 228, 230, 234, 243, 248, 252, 257, 261, 262, 265, 268, 270, 271, 273, 274, 276–287, 292, 293, 295–297, 299, 302–305, 307–314, 316, 325–327
  acoustic isolation, function of, 3, 35, 101
  adhesive properties of, 164, 166, 168, 238, 282, 283, 296, 304

Gaskets (*Cont.*):
  antistick properties of, 164
  ASTM test methods of, 10, 16
  beater addition, 5, 25, 27, 30, 32, 34
  breaking strength of, 33, 139
  chemically compatible, 105
  compressibility of, 6, 7, 10, 16, 32, 34,
    38, 41, 47, 52, 67–70, 179, 194, 230,
    232, 282, 283, 305, 313, 314
    test method for, 6, 7, 10–23, 70, 117,
      127, 128, 130, 135, 136, 148, 150,
      169, 173, 195, 199, 221–224,
      227–234, 246, 325, 326
  compression curve for, 68, 72, 75
  compressive strength of, 35, 36, 38
  configuration of, 1, 47, 52, 56, 83, 184,
    195, 213, 214, 216
  conformability of, 34, 44, 55, 66, 67,
    203
  corrosion testing of, 10, 14
  creep of, 12, 16, 20, 36, 52, 68, 89,
    132–134, 136, 141, 142, 146, 149,
    191, 206, 208, 209, 221–223,
    231–233, 268, 303
  creep-relaxation of:
    test method for, 6, 7, 10–23, 70, 117,
      127, 128, 130, 135, 136, 148, 150,
      169, 173, 195, 199, 221–224,
      227–234, 246, 325, 326
  definition, 1, 89, 139, 142, 146, 209,
    234, 257, 267
  density, 27, 32, 36, 41, 46, 47, 55, 77,
    78, 91, 141, 142, 191, 256, 270, 280,
    313, 327
  design, 32, 49, 51, 55, 67, 69, 79, 82, 84,
    104, 105, 114, 125, 128, 131, 150,
    191, 247, 253, 254, 257, 260, 287,
    290, 292, 293, 297, 300, 305, 307,
    315, 325
  dimensional stability of, 10, 16, 35
  erosion of, 203, 300, 303
  eyeletting of, 164, 243, 297, 309
  fabrication, 271
  fastener for, 55, 56, 66, 245, 250, 278,
    290
    bolt holes, 58, 61, 159, 164, 187, 284,
      285, 309, 312
      pattern of, 53, 59, 60, 62, 63, 65, 78,
        81, 185, 197, 245, 262
      tightening of, 59, 63, 76, 100, 125,
        326
  fibers and fillers, 29, 32
  flange bowing and, 54

Gaskets (*Cont.*):
  flanges for, 1–3, 14, 34–36, 41, 44, 45,
    48, 51, 55–59, 61, 63–66, 68, 69,
    73–75, 78, 90, 105, 128, 130, 150,
    153, 158, 159, 163, 164, 176, 177,
    183, 210, 223, 229, 230, 232, 234,
    246, 252, 276–278, 282, 283,
    285–287, 289, 303, 305, 312, 314
  fluid resistance of, 13, 18, 32, 37, 39,
    227
  gasket and joint diagram, 153
  gasket behavior at elevated temperature, 133
  installation of, 34, 114, 158, 187, 218,
    219, 249, 250, 283, 312, 320
  leakage rate in, 223, 224, 226, 229, 265,
    266, 268
  load compression curve for, 75
  manufacturing of, 25, 144, 191, 211,
    217, 253, 254, 256, 257, 260, 326
  material characteristics, 262
  material combinations, 1, 3, 4, 11, 24,
    41, 46, 52, 56, 59, 89, 141, 145, 153,
    184, 235, 260
  metallic, 42, 44–46, 64, 68, 100, 135,
    144, 163, 166, 230, 237, 271, 304,
    313, 320
    spring rate of, 44, 93, 95, 155, 156,
      186, 232
  MTI:
    test procedures for, 7, 23, 25, 127,
      130, 136, 138–142, 145, 146, 150
  printing on, 33, 163, 180, 185, 276, 305
  PVRC test procedures for, 23, 25, 105,
    117, 120, 122, 123, 127–132, 136,
    137, 142, 148–151
  radial strength of, 178
  recovery of, 3, 12, 16, 31, 34, 42, 46,
    68, 70, 90, 180, 182, 184, 186,
    190, 192, 203, 208, 230, 232, 277,
    308
  relaxation curve, 12, 23, 24, 36, 38, 46,
    47, 52, 53, 68, 89–91, 93, 96–98,
    130–132, 134, 136, 139, 141, 142,
    146–149, 154, 174–176, 179, 191,
    205, 206, 208, 209, 217, 218,
    221–223, 231, 233, 243, 245, 248,
    303
  scuff resistance of, 33
  sealing performance of, 2, 24, 33–35,
    54, 91, 151, 248, 263, 316–318
  seating stresses for, 53, 111
  shape factor, 96–98, 194, 208

332 Index

Gaskets (Cont.):
  shear strength of, 20, 35, 56, 190–192, 194, 195, 206, 208, 209
  spring rate of, 44, 93, 95, 155, 156, 186, 232
  stress compression for, 67, 154, 155, 241
  stress distribution in, 69, 70, 77–79, 84, 105, 180, 182, 193
  stress relaxation for, 89, 132, 175, 206, 209
  surface finish, 3, 34, 44, 46–48, 51, 52, 54, 55, 64, 66, 111, 113, 219, 245, 287, 316
  tensile strength of, 14, 17, 32, 37, 39, 48, 134, 136, 138–141, 190, 197, 207, 209, 230
  thermal analysis of, 235, 236
  thermal conductivity of, 10, 65, 196, 239, 289
    test method for, 6, 7, 10–23, 70, 117, 127, 128, 130, 135, 136, 148, 150, 169, 173, 195, 199, 221–224, 227–234, 246, 325, 326
  thickness of, 3, 6, 27, 29, 32, 35, 36, 45–47, 54–57, 68–70, 72–76, 82, 90, 91, 96, 97, 166, 173–176, 178, 179, 184, 185, 189, 191, 192, 194, 196, 206, 208, 209, 230, 234, 250, 256, 273, 274, 277, 278, 281, 289, 298, 299, 303–305, 307
  tightness parameter measurement of, 24, 118, 120, 142
  weight loss in, 15, 24, 134, 138, 141, 142, 146, 147
  width of, 46, 47, 56, 60, 76, 98, 108, 111, 115, 186, 187, 285, 299
Graphite gaskets, 3, 5, 25, 29, 30, 32, 34, 41, 46, 134, 135, 141, 142, 144, 146–148, 151, 305, 310, 325, 327

Hardness, 10
Heat age, 10
Heat conductivity, 35
Heat resistance, 10, 19, 34, 36–39, 44, 207, 221, 227, 282, 303
Hydrostatic end force, 53, 84, 90, 107, 111, 113, 114, 153, 292

Ignition loss, 10
Infrared spectroscopy, 238

Initial seal creation, 66, 67
Injection molding, 200
Intake and exhaust manifold gaskets, 308
Internal combustion engine gaskets, 287, 325

Jacketed gaskets, 135
Joint design, 2, 48, 87, 117, 131, 148, 150, 325

Laser cutting, 271, 272
Lead pellet testing, 72
Leak rate, 10, 15, 24, 118, 120, 143
Leak rates versus "y" stresses, 10
Low temperature, 10

M factor, 15, 107, 113, 115, 229
Machined metal gaskets, 48
Maintenance of the seal, 89
Material characteristics for processing, 32
Material Test Institute/Pressure Vessel Research Institute test procedures, 7
Materials Technology Institute (MTI), 23
Mechanical test (HOMT), 133
Metal wire, 48
Mica, 41, 163, 164
Micro-conformability, 34
Midpoint loading, 57, 60
Molded gaskets, 202
Molded rubber gaskets, 203, 210
Molybdenum disulfide, 163, 310

Natural, 39
Natural rubber (NR), 39, 40, 197
Nitrile (Buna N), 39
Nonmetallic gaskets, 2, 42

O-ring:
  diametral clearance, 216
  extrusion, 31, 32, 35–39, 56, 174, 177, 213, 216–219, 222, 234, 303
  grooves, 42, 211, 219, 300
  military standard, 210
  squeeze, 28, 90, 215, 219
  swell, 29, 31, 40, 65, 216
Ozone, 10

Index

Peforated core, 174, 176, 178–180, 296
Plastics, 42
Polyacrylic, 41, 164, 185, 303
Polycloroprene, 39
Polyurethane, 39
Pressure Vessel Research Council (PVRC), 23, 105, 127, 136
Printing, 163
Pulsator, 243, 250

Quality parameters, 138–140, 142, 144, 147
Quantification of aging effects, 135

Radial strength, 35, 178
Recovery, 7
Regular carbon paper, 70, 77, 78
Reinforced gaskets, 142, 168
Resilience, 10
Resistance, 10
Ridge and groove gasket, 300
RTV (room temperature vulcanizing), 276–284, 300
Rubber, 4, 5, 11, 13, 14, 17–23, 25, 28–32, 36, 37, 39–41, 53, 54, 65, 185, 189–199, 202, 203, 206, 208, 210, 223, 229, 233, 238, 239, 261, 271, 274, 276, 277, 281, 296, 297, 303, 305, 307–309, 311–313
  abrasion resistance, 192
  adhesion, 10, 18, 24, 130, 195, 231, 277, 278, 282, 314
  aging, 24, 25, 27, 29, 133–148, 186, 191, 197, 205, 208, 209, 237, 303
  ASTM Test Methods for vulcanized elastomers, 10, 16
  chemical resistance, 38, 41, 160, 198, 199, 282
  compression, 12, 13, 16, 19, 20, 40, 44, 46, 47, 52, 53, 55, 56, 58, 67, 68, 72–77, 89, 93, 95, 107, 132, 134, 135, 137, 153–157, 174, 184, 186, 189–192, 194, 200, 202, 206, 208, 209, 211, 219, 222, 232, 234, 240–242, 285, 289, 290, 299, 304, 305
  compression molding, 200
  compression set, 12, 16, 40, 56, 156, 157, 191, 222, 232, 234, 285
  corona, 198

Rubber (Cont.):
  creep relaxation, 12, 36, 191, 221, 223, 231, 233
  electric properties, 195
  flex fatigue, 193
  hardness, 23, 190, 191, 207, 217, 274
  heat buildup, 192
  heat resistance, 19, 34, 36–39, 44, 207, 221, 227, 282, 303
  injection molding, 200
  low temperature properties, 208
  modulus, 20, 54, 190, 191, 194, 196, 205–209, 217, 236
  ozone deterioration, 198
  permanent set, 42, 132, 192, 208
  permeability, 68, 186, 193
  resilience, 20, 21, 45–47, 143, 191, 192, 194, 206, 208
  shape factor, 96, 97, 98, 194, 208
  strength, 14, 17, 19, 27, 31–33, 35–39, 48, 134, 136, 138–141, 166, 168, 178, 190, 193, 195, 197, 207, 209, 230, 277, 278, 282, 283, 285, 289, 304
  tear strength, 19, 193
  tensile, 14, 17, 23, 24, 31, 32, 34, 35, 37, 39, 48, 77, 99, 130, 134, 136, 138–141, 190, 191, 197, 205–209, 230, 231, 249
  thermal properties, 196
  transfer molding, 200
  water resistance, 198
Rubber Glossary:
  abrasion, 66, 192, 193, 203
  accelerator, 203
  aging, 24, 25, 27, 29, 133–148, 186, 191, 197, 205, 208, 209, 237, 303
  antioxidant, 205
  average room conditions, 205, 209
  bench marks, 205
  brittle point, 205, 207
  calender, 27, 74, 205
  cold flow, 46, 205
  cold resistance, 205
  compression set, 12, 16, 40, 56, 156, 157, 191, 222, 232, 234, 285
  concavity factor, 206
  conditioning, 74, 76, 173, 206
  creep, 12, 16, 20, 36, 52, 68, 89, 132–134, 136, 141, 142, 146, 149, 191, 206, 208, 209, 221–223, 231–233, 268, 303
  crystalization, 22, 197, 206, 236

Rubber (*Cont.*):
  cure, 3, 190, 191, 197, 206, 276, 278–282, 284–286
  damping, 192, 206, 237
  deformation, 20, 53, 56, 100, 189, 191, 192, 194, 206–209, 216, 217, 237, 268
  diluent, 206
  drift, 89, 206
  dynamic fatigue, 18, 206
  dynamic modulus, 20, 194, 206
  dynamic resilience, 206
  elongation, 17, 90, 100, 154, 190, 191, 206, 208–210, 234, 245, 277
  filler, 39, 41, 46, 135, 196, 207, 208, 238, 295
  flex resistance, 18, 207
  flexing, 20, 196, 207
  freezing point, 190, 207, 210
  gage length, 207, 209
  growth, 65, 207, 302, 320
  hardness, 23, 190, 191, 207, 217, 274
  heat build up, 192
  heat resistance, 19, 34, 36–39, 44, 207, 221, 227, 282, 303
  hot tensile, 207
  hysteresis, 132, 192, 194, 203, 206, 207
  immediate set, 207
  inert filler or pigment, 207
  Joule effect, 196
  memory, 189, 207
  modulus, 20, 54, 190, 191, 194, 196, 205–209, 217, 236
  mold, 65, 185, 196, 200, 208, 260
  oil resistance, 199, 208
  permanent set, 42, 132, 192, 208
  plasticizer, 208, 219, 237, 238
  proof resilience, 206, 208
  reinforcing pigment or agent, 208
  relaxation, 12, 23, 24, 36, 38, 46, 47, 52, 53, 68, 89–91, 93, 96–98, 130, 131, 132, 134, 136, 139, 141, 142, 146–149, 154, 174–176, 179, 191, 205, 206, 208, 209, 217, 218, 221–223, 231, 233, 243, 245, 248, 303
  relaxation time, 208
  resilience, 20, 21, 45–47, 143, 191, 192, 194, 206, 208
  rigidity, 34, 35, 42, 44, 51, 52, 56, 69, 133, 134, 136, 191, 208, 226, 287

Rubber (*Cont.*):
  set, 16, 20, 40, 42, 53, 56, 132, 137, 145, 156, 157, 176, 177, 186, 191, 192, 203, 205–209, 211, 222, 231, 232, 234, 237, 248, 285, 305
  set at break, 208
  shape factor, 96–98, 194, 208
  shear modulus, 191, 208
  spring constant, 132, 208
  static fatigue, 209
  static modulus, 20, 194, 209
  stiffness, 132, 157, 191, 196, 197, 207–210, 226, 236, 312
  strain, 12, 89–100, 189, 190, 192, 194, 196, 205–209, 229, 233, 242, 249, 250, 290
  strain relaxation, 205, 209
  stress, 12, 16, 17, 44, 51–53, 57, 62, 66, 67, 69, 70, 72, 77–79, 82–84, 89–91, 93, 95, 97–99, 105, 107, 108, 111, 113, 114, 117, 118, 120, 122, 124, 128, 132, 133, 135–138, 142, 149, 154–156, 164, 174–177, 179, 180, 182, 190, 192–195, 198, 206–209, 217, 218, 227, 228, 232, 233, 235, 236, 240, 241, 245, 249, 262, 277, 290, 292, 296, 299, 302–305
  stress relaxation, 89, 132, 175, 206, 209
  subpermanent set, 209
  tangent modulus, 209
  tear resistance, 19, 193, 209
  tensile product, 206, 209
  tensile pull, 34
  tensile strength, 14, 17, 32, 37, 39, 48, 134, 136, 138–141, 190, 197, 207, 209, 230
  tensile stress, 17, 207, 209
  transition points, 205, 207, 210
  ultimate elongation, 210
  vulcanization, 18, 29, 194, 203, 206, 208, 210

Sandwich gasket, 295
Saturating, 160
Scanning electron microscopy, 240
Scuff resistance, 33
Sealability, 7, 12, 34–37, 55, 67, 163, 164, 174, 177, 221–224, 226–228, 303, 318, 319
Sealants, 3, 283, 286
Segmented gaskets, 186

Shear strength, 20, 35, 56, 190–192, 194, 195, 206, 208, 209
Silicone, 39, 40, 41, 163, 182, 185, 261, 276–284, 300, 305, 312
Simplified procedure, 111
Simplified procedure proposed by Whalen, 105
Spiral wound gaskets, 47, 143
Spring rate, 44, 93, 95, 155, 156, 186, 232
Standardized leak rate, 120
Steam test, 148, 249, 250
Steel rule dies, 271
Stress compression, 67, 154, 155, 241
Stress distribution, 69, 70, 77–79, 84, 105, 180, 182, 193
Stress relaxation, 89, 132, 175, 206, 209
Styrene Butadione (Buna S), 39
Surface finish, 3, 34, 35, 44, 46–48, 51, 52, 54, 55, 64, 66, 111, 113, 219, 245, 287, 316

Tanned glue and glycerin, 41
Tear strength, 10
Teflon, 42, 46, 163, 218, 305
Tensile strength, 14, 17, 32, 37, 39, 48, 134, 136, 138–141, 190, 197, 207, 209, 230
Tension loss, 100, 101
Tension testing, 10
Thermal:
  analysis (TA), 235, 236

Thermal (*Cont.*):
  conductivity, 10, 65, 196, 239, 289
  gravimetric analysis (Tga), 237
  mechanical analysis (Tms), 237
Thermal conductivity, 10
Tightness parameter, 24, 118, 120, 142
Tool wear and life, 33
Torque, 3, 52, 62, 63, 66, 68, 74, 76, 78, 90, 95, 98–101, 103, 104, 108, 114, 159, 176, 178, 179, 210, 222, 234, 263, 278, 283–285, 296, 303, 304, 305, 313, 314, 325
Torque loss, 3, 68, 95, 98–101, 178, 179, 210, 222, 234, 278, 285, 313, 314

Ultrasonic instrument, 243
Ultraviolet (Uv), 65, 270
Unbroken metal core, 166, 168, 174, 176, 179, 180, 303

Viscoelastic gasket, 52
Viton, 39
Volume control rubber gasket, 202

Water jet cutting, 272
Wire ring, 299

Y factor, 107, 122

About the Author

Daniel E. Czernik is Vice-President of Advanced Technology for Fel-Pro, Inc. in Skokie, Illinois, where he has been employed since 1969. Mr. Czernik holds numerous patents for sealing products, and he has written many influential papers in the field of sealing technology. He is a Fellow Member of the Society of Automotive Engineers. He is also the coauthor of *Handbook of Fluid Sealing,* published by McGraw-Hill.